Counterinsurgency Warfare and Brutalisation

This book offers the first analysis of the brutalisation paradigm in counterinsurgency warfare.

Minimising the use of force and winning over the population's opinion is said to be the cornerstone of success in modern counterinsurgency (COIN). Yet, this tells only one side of the story. Drawing upon primary data collected during interviews with eyewitnesses of the Second Russian-Chechen War, as well as from secondary sources, to the best of the authors' knowledge, this book is the first to offer a detailed analysis of the long-neglected logic underpinning brutalisation-centred COIN campaigns. It offers a comprehensive systematisation of the brutalisation paradigm and challenges the widespread assumption of brutalisation as an underperforming paradigm of COIN warfare. It shows that, although appalling, brutalisation-centred measures can deliver success. This book also outlines a stigmatised yet widely deployed set of COIN measures and provides critical insights into how Western military blueprints can be improved without compromising important moral and ethical requirements.

This book will be of much interest to students of counterinsurgency, military and strategic studies, Russian politics, and International Relations.

Roberto Colombo is an independent researcher and security analyst specialising in counterinsurgency and counterterrorism.

Emil Aslan Souleimanov is a professor at the Institute of Political Studies, Faculty of Social Sciences, Charles University, Czech Republic.

Cass Military Studies

Understanding Insurgent Resilience
Organizational Structures and the Implications for Counterinsurgency
Andrew D. Henshaw

Military Strategy of Middle Powers
Competing for Security, Influence and Status in the 21st Century
Håkan Edström & Jacob Westberg

Military Mission Formations and Hybrid Wars
New Sociological Perspectives
Edited by Thomas Vladimir Brond, Uzi Ben-Shalom and Eyal Ben-Ari

French Defence Policy Since the End of the Cold War
Alice Pannier and Olivier Schmitt

War and International Relations
A Critical Analysis
Balazs Szanto

A Global History of Pre-Modern Warfare
Before the Rise of the West, 10,000 BCE—1500 CE
Kaushik Roy

Military Strategy of Great Powers
Managing Power Asymmetry and Structural Change in the 21st Century
Håkan Edström and Jacob Westberg

Counterinsurgency Warfare and Brutalisation
The Second Russian-Chechen War
Roberto Colombo and Emil Aslan Souleimanov

For more information about this series, please visit: www.routledge.com/
Cass-Military-Studies/book-series/CMS

Counterinsurgency Warfare and Brutalisation
The Second Russian-Chechen War

**Roberto Colombo and
Emil Aslan Souleimanov**

Routledge
Taylor & Francis Group

LONDON AND NEW YORK

First published 2022
by Routledge
2 Park Square, Milton Park, Abingdon, Oxon OX14 4RN

and by Routledge
605 Third Avenue, New York, NY 10158

Routledge is an imprint of the Taylor & Francis Group, an informa business

© 2022 Roberto Colombo and Emil Aslan Souleimanov

British Library Cataloguing-in-Publication Data
A catalogue record for this book is available from the British Library

Library of Congress Cataloging-in-Publication Data
A catalog record for this book has been requested

ISBN: 978-1-032-03579-6 (hbk)
ISBN: 978-1-032-03581-9 (pbk)
ISBN: 978-1-003-18804-9 (ebk)

DOI: 10.4324/9781003188049

Typeset in Times New Roman
by Apex CoVantage, LLC

Contents

Figures

Tables

Preface

Minimising the use of force and winning over the population's hearts and minds are said to be the cornerstones of success in modern counterinsurgency (COIN). Yet, this is only one side of the coin. Drawing upon secondary sources and primary data collected during face-to-face interviews with eyewitnesses of the Second Russian-Chechen War, to the best of the authors' knowledge, this book is the first to offer a detailed analysis of the long-neglected logic underpinning brutalisation-centred COIN campaigns. Venturing into an underresearched area, it provides unique contributions to the existing literature. First, this book offers a comprehensive systematisation of the brutalisation paradigm, so far lacking in the literature advancing a critical reading of COIN warfare. Second, it challenges the widespread assumption of brutalisation as an underperforming paradigm of COIN warfare and shows that, although appalling, brutalisation-centred measures can deliver success. Third, it sheds light on a stigmatised yet widely deployed set of COIN measures. Spearheading this alternative stream of policy-oriented research, it provides critical insights into how Western military blueprints can be improved without compromising important moral and ethical requirements. Fourth, empirically, it offers an original revisitation of the Second Russian-Chechen War. This book calls for future research into the brutalisation paradigm.

Acknowledgements

Research on the present manuscript was enabled by a research grant provided by the Grant Agency of the Czech Republic 21–14872S: "Fratricidal defection: How blood revenge shapes anti-jihadist mobilisation."

Abbreviations

AP I	First Additional Protocol to the Geneva Conventions
ChRI	Chechen Republic of Ichkeria
COIN	Counterinsurgency
DoD	Department of Defence
FLN	National Liberation Front
FM 3–24	Field Manual 3–24
HUMINT	Human Intelligence
IHL	International Humanitarian Law
IO	Information Operation
KHAD	Khadamat-e Aetla'at-e Dawlati
LOAC	Law of Armed Conflict
NATO	North Atlantic Treaty Organisation
PRT	Provincial Reconstruction Team
ROE	Rules of Engagement
SIGAR	Special Inspector General for Afghanistan Reconstruction
SS	Schutzstaffel
VCI	Viet Cong Infrastructure

1 Introduction

We used tough methods to show what's wrong and what's right. Against those who didn't understand, we led a tough and even cruel struggle.
—Ramzan Kadyrov, Head of the Chechen Republic
(quoted in Stack 2008)

According to Western military doctrines, promoting good governance and winning the population's outright support constitute the hallmarks of a perfect counterinsurgency (COIN): "dollars and ballots may have more important effects than bombs and bullets" (U.S. Gov. 2014: 7–2). Yet, this "infallible" theory of COIN warfare centred on minimising the use of force and creating a secure environment for the population has often failed to deliver impressive results. The peace agreement signed on 29 February 2020 between the United States (U.S.) and the Afghan Taliban represents the latest attempt of a fatigued nation to put an end to America's longest war. "We are seizing the best opportunity for peace in a generation," explained Secretary of State Mike Pompeo after the signing ceremony, acknowledging that the Talibans have been successfully holding ground against one of the world's most powerful militaries for almost two decades (U.S. Gov 2020). Ten years earlier, in February 2008, the then-prime minister of the Russian Federation Vladimir Putin used different words to describe what victory against the Chechen insurgency looked like: "we have dealt it a decisive and crushing blow" (Putin 2008). On its way to success in Chechnya's inhumane civil war, Russia has followed a strategic blueprint that could not be more inimical to the one proposed in Western military textbooks. The Russians waged a campaign of indiscriminate shelling on populated settlements, tortured and blackmailed people to collect intelligence, and engaged in the deliberate victimisation of the Chechen population to frighten entire communities into passive submission (Schaefer 2010).

DOI: 10.4324/9781003188049-1

How is it possible for an incumbent to suppress a rebellion in spite of—or thanks to—its complete disregard for the most fundamental tenets of modern COIN warfare? Recent research on the drivers of success in asymmetric conflicts indicate that gaining the population's trust, minimising the use of kinetic force, and solving popular grievances are the cornerstones of victory, while a growing literature stresses the importance of abiding by ethical and moral prescriptions during the conduct of COIN operations (Mockaitis 2003; Nagl 2005; Lavine 2010; Dowdall & Smith 2010). If prevailing against insurgency is contingent upon the strict adherence to the guidelines advanced in Western military doctrines, then it is hard to explain the trail of successes achieved by counterinsurgents resorting to indiscriminate killings, mass torture, and other ruthless techniques of warfare (Zhukov 2008, 2011). Research into the logic of brutalisation-centred approaches is surprisingly underdeveloped and suffers from the widespread assumption that deploying large-scale coercion irremediably depletes a counterinsurgent's chances of success. Little is known about why, when, and how counterinsurgents resorting to the brutalisation paradigm can achieve victory, nor is it clear why brutalisation should be an inherently poor COIN concept. Brutalisation-centred measures have found widespread application in past and present theatres of COIN warfare, yet their controversial nature has discouraged the scholarship from venturing into this long-ignored variant of COIN warfare. Experts advancing a critical reading of COIN campaigns have launched all-out attacks against the fundamentals of the hearts and minds approach, without however paying much attention to the relevance of their findings for contemporary practitioners. It follows that Western decision-makers are aware of the hearts and minds approach's limitations but are unfamiliar with the alternatives available to counterinsurgents unable to garner popular support.

We contribute to fill this gap in the existing literature by conducting an in-depth inquiry into the brutalisation paradigm. Breaking the intellectual impasse affecting the study of controversial pathways of COIN warfare, we advance this alternative stream of policy-oriented research and offer an unprecedented inquiry into a long-neglected yet widely deployed set of COIN techniques. We draw on extensive empirical data and engage in theoretical debates to systematise the ways in which heavy-handed intelligence, information, military, political, and economic measures assist counterinsurgents in defeating grassroots rebellions. The rigorous assessment of the available data on the Second Russian-Chechen War of 1999–2009 and other theatres of COIN warfare reveals that the brutalisation paradigm can be as effective as, if not more effective than the hearts and minds approach when there is little to no chances of winning over the population's support. These findings hold enormous relevance for scholars of security studies and strategic decision-makers. Fighting insurgency is an extremely complex endeavour, and decision-makers are required to constantly re-assess the results obtained in past and present experiences if they want to improve and optimise the strategic toolkits at their disposal. As waging COIN warfare on the ground can look substantially different from the polished accounts provided in military manuals, a counterinsurgent should be aware of which techniques can and cannot work in a given context. Minimising the use of force and promoting good governance are desirable and noble aims, but if doing so does not deliver satisfactory results, a counterinsurgent should be ready to consider alternative options. While this book strongly discourages Western decision-makers from engaging in endeavours that put the dignity and integrity of human life at risk, its findings underscore that the brutalisation paradigm offers a controversial yet effective pathway when other, preferable measures fail in achieving their intended purposes.

Aims and scope

The current discourse on modern COIN warfare is dominated by a single paradigm, permeating the doctrinal guidelines of Western counterinsurgents, known in military circles as the "hearts and minds" approach. At first sight, this is not an issue. Waging COIN warfare from a position of moral authority and with sincere concern for the local populations' safety could not be better aligned with the principles of humanity, proportionality, and discrimination followed by Western military forces (U.S. Gov. 2016). Yet, waging an ethical COIN warfare does not come free of charge. In the past, counterinsurgents that failed to gain the local population's support have suppressed grassroots rebellions by resorting to their military prowess instead. This was the case of the Roman Empire, capable of keeping control over its outer regions with an iron fist: "They make a desolation, and they call it peace" (Tacitus 1914[98 AD]: 221). But because today's counterinsurgents cannot do as the Romans do, many scholars and practitioners accept that modern-day COIN operations can rarely deliver a full-fledged victory: "messy and unsatisfying are the hallmarks of success in modern counterinsurgency wars" (Nagl 2012).

This is not to say that the hearts and minds approach cannot be successful. On the contrary, addressing the root causes that gave rise to the insurgency whilst providing security and economic assistance to the local population can contribute to permanently extinguishing the "cinders" of discontent that insurgent movements exploit to bring rebellions back to life (Glenn 2015: 14). Yet, succeeding in fighting back the insurgents, providing security to the population, rebuilding a war-torn country anew, and at the same time restoring legitimacy for the central government is a "dauntingly complex task," which often sits beyond a counterinsurgent's grasp (Roxborough 2007: 18). Counterinsurgents who succeed in these endeavours might be rewarded with the attainment of long-lasting peace in a post-conflict scenario. However, achieving victory by winning over the population's hearts and minds can often be a long, difficult, and exhausting journey. Given the inherent complexities of waging a hearts and minds operation, one might ask whether alternative pathways can promise faster results at lower political, economic, and military costs.

The scholarship's scepticism towards the hearts and minds approach found fertile ground in the unsatisfying results recently achieved by Western counterinsurgents. Bogged down in the quagmires of Afghanistan and Iraq, the hearts and minds approach has lost much of its former glory. Some commenters manifested their mistrust towards COIN warfare, arguing that the hearts and minds approach is rarely if ever successful: "Counterinsurgency was a recipe for defeat and retrenchment in the recent past, just as it was in the 1970s and will be again" (Jeffrey 2015: 180). Others criticised Western decision-makers for their lack of strategic foresight, arguing that a kinder way of COIN warfare can only exist in the minds of those who never experienced war in the first place: "Our civilian and uniformed leaders have engaged in comforting fantasies about the multi-layered conflicts we're in, while speaking in numbing platitudes. Now we're back to 'winning hearts and minds.' We can't do it." (Peters 2006). Given these criticisms, it is surprising to see that strategic blueprints based on promoting good governance and winning the population's support *are not* undergoing a substantial re-evaluation. This has likely not been done because human security dogmas are so firmly entrenched in the strategic approach of Western states that, if there is a chance for achieving victory, it can only be through winning over the population's hearts and minds.

Yet, hearts and minds are not the only game in town. And, they are not the only approach that counterinsurgents dispose of confront and defeat grassroots rebellions. Next to the hearts and minds approach lies a darker way of performing COIN warfare, referred to in this book as the brutalisation paradigm. Casted at the margins of the COIN literature, this approach has gained a reputation for being a checklist of everything there is of immoral, repugnant, and counterproductive in a COIN operation (Paul & Clarke 2016). Unappealing to Western audiences and rebuffed in scholarly works, an in-depth understanding of brutalisation and its logic is lacking. No study has, to date, provided a comprehensive analysis of brutalisation-centred COIN campaigns, nor do we possess a solid understanding of how the brutalisation paradigm brings about success—if it can deliver victory at all.

Challenging the widespread assumption of the hearts and minds approach as a universal recipe for COIN warfare, this book is the first to advance a comprehensive analysis of the brutalisation paradigm as an alternative pathway to mission success. Acknowledging brutalisation as a set of ethically questionable measures, this book identifies and critically assesses the components of a successful brutalisation-centred COIN campaign. It illustrates the ways in which counterinsurgents deploy heavy-handed intelligence, information, military, political, and economic means to isolate, contain, and ultimately suppress an insurgency. As the logic behind brutalisation-centred measures remains nebulous, strengthening our understanding of alternative approaches to COIN warfare assumes a role of primary importance. This book systematises our fragmentary knowledge of the brutalisation paradigm and revisits widespread inaccuracies on the role of coercion in COIN warfare. It shows that the brutalisation paradigm might be more effective than its "kinder, gentler" counterpart in breaking an insurgency's backbone when winning over the population's hearts and minds is either unfeasible or outright impossible. Empirically, this book offers an in-depth revisitation of a widely misinterpreted case of a brutalisation-centred COIN campaign, namely the COIN operations performed by Russia during the Second Russian-Chechen War of 1999–2009.

These observations should not mislead the reader into assuming that Western policy-makers would find it appropriate to adopt the full spectrum of measures associated with the brutalisation paradigm. During a COIN campaign, Western states must abide by domestic laws and international treaties that regulate the use of armed force to prevent and contain the incidence of collateral damage. Although states adopt different legal frameworks according to the conventions and treaties that they choose to sign, two principles regulating the conduct of military hostilities have gained the rank of customary law. Listed as a "basic rule" of armed conflicts in the First Additional Protocol to the Geneva Conventions relating to the protection of victims in armed conflicts (hereinafter AP I), the principle of distinction binds belligerents to distinguish at all times "between the civilian population and combatants and between civilian objects and military objectives" (ICRC 1977: art. 48). Nevertheless, in asymmetric conflicts, the counterinsurgent is often prevented from accurately discerning insurgent fighters from civilian non-combatants. To compensate for the limitations that the principle of distinction bears in the context of COIN warfare, counterinsurgents are required to scrupulously follow the principle of proportionality, which prevents states from launching attacks that "may be expected to cause incidental loss of civilian life . . . which would be excessive in relation to the concrete and direct military advantage anticipated" (Ibid.: art. 51, 5[b]; McLeod 2014: 68). As explained by Kahl, a military action can only take place when the counterinsurgent fully satisfies both principles: "if both the distinction and proportionality principles are met, civilian casualties resulting from a strike on a military target, however tragic, are not considered violations of international law" (Kahl 2007: 10).

It is within the limits set forth by International Humanitarian Law [IHL] principles that this book draws from the brutalisation paradigm a list of takeaways for Western decision-makers. As incumbents waging brutalisation-centred COIN campaigns deliberately and systematically breach the principles of distinction and proportionality, advocating in favour of such measures would be not only morally appalling but also unacceptable under any domestic or international legislation. As the *U.S. Field Manual 3–24* (hereinafter *FM 3–24*) explicitly states, "under no circumstances may military necessity authorise actions specifically prohibited by the law of war" (U.S. Gov. 2014: 13-3). While diving deep into overlooked variants of COIN warfare enriches our understanding of what works and what does not in a given operational theatre, neglecting the importance of IHL principles would risk compromising the effectiveness of COIN operations performed according to the hearts and minds approach.

Data and methods

To provide a high degree of contextual insight to the brutalisation paradigm, this book adopts a single case-study research design (Dyer & Wilkins 1991). We select the Second Russian-Chechen War as a case study because many of the brutalisation-centred measures utilised by contemporary counterinsurgents were tested and perfected by the Russians in Chechnya. As sustained by Blank (2016: 81), the Russian COIN in Chechnya is "the latest adaptation" of a "well-established history" of brutalisation-centred measures that find their most recent operational battleground in the ongoing Syrian Civil War (Haines 2016; Avramov 2018). While the empirical analysis focuses on the Russian COIN operations in Chechnya, this book draws extensively from a wide array of additional case studies to give empirical flesh to its findings.

Empirically, this book draws upon an extensive analysis of the literature available on the Second Russian-Chechen War, which is integrated with a sample of eight exclusive semi-structured interviews carried out with Chechen refugees in the region of the Pankisi Gorge, Georgia. Participants have been recruited following a "snowballing sampling" procedure, a non-random sampling technique appropriate for accessing members of "unique, hard-to-reach, or marginalised populations" (Tenzek 2018: 1614). By exploiting the tight social interconnectedness typical of rural, isolated village communities, this sampling method allowed the researchers to access a target audience otherwise impossible to reach. When referring to interviewees, this book utilises alphanumeric codes, concealing names, genders, and professions to protect the participants' identity. As the Pankisi Gorge is a demographically and geographically small area, providing additional information on the participants' identities would risk infringing essential anonymity requirements. It is however specified that two of the eight interviewees were former insurgents, as their accounts provide a "military spin" to the data collected, complementing and reinforcing the more civilian-centred perspective provided by the other participants.

The Chechen community of the Pankisi Gorge provides an optimal pool of potential interviewees for three main reasons. First, most of the Chechens living in the Pankisi Gorge maintain regular contact with their relatives in Chechnya. Due to the proximity of the region to the Chechen territory, many residents frequently receive guests and/or visit their families across the border. Thus, the interviewees not only could provide first-hand accounts of the war but were also in the position to discuss the long-term effects of the Russian COIN operations. Second, the possibility of conducting interviews in Chechnya is very limited. While Human Rights Watch reported in 2016 that local villagers would refuse to interact with its personnel due to fear of triggering the government's reprisals (Human Rights Watch 2016), Iliyasov recently confirmed that the climate of fear enforced by the regime "prevents people from participation in any kind of interview" (Iliyasov 2019: 1710). Third, the quality of information collected from people living in Chechnya can often be of poor quality, as interviewees might deliberately provide inaccurate or deceptive information to protect their families from the regime's retaliation. The "chilling effect" produced on the data collected in Chechnya has been well documented by Ratelle, who reports that most of his interviewees provided standardised answers that added little to no novel information for research (Ratelle 2013: 219–220). This was confirmed by one interviewee, who stated that "to this day, people cannot speak freely about the situation in Chechnya. I can, but only because I am in Georgia."[1] Interviews were audio-recorded with the participants' consent and lasted 40 minutes on average.

To process the data collected, this book follows a thematic analysis procedure, defined by Maguire and Delahunt as the "process of identifying patterns or themes within qualitative data" (Maguire & Delahunt 2017: 3352). The analysis includes the content extrapolated from the interviewees' direct statements (manifest content) as well as the underlying aspects of the emerging patterns (latent content) (Boyatzis 1998: 16). While a manifest analysis is necessary to identify significant thematic areas, a latent approach allows the researcher to disclose reasonings left unexpressed or implied in the conversation. For instance, a latent analysis is necessary to interpret the passages in which interviewees assess the economic conditions of their fellow countrymen living in Chechnya. As statements such as "in comparison with my relatives . . . I am very poor"[2] and "when I go [to Chechnya], it is like going from rags to riches"[3] are symptomatic of grievances felt by the Chechens in Georgia, but that do not necessarily express the views of individuals living in Chechnya, a latent analysis limits the contamination of data and enables the accurate triangulation of the information obtained through a manifest analysis.

Given the relatively small pool of interviewees, we are aware that the primary sources advanced in this book do not constitute a representative sample. One could argue that additional interviewees would have better consolidated this research's findings or potentially disclosed novel patterns in the participants' responses. As these are reasonable concerns, all necessary precautions were taken during the data collection and analysis to minimise the incidence of participant bias. While the use of a diversified range of probing questions was specifically aimed at preventing the naïve incorporation of group narratives in the research output (Franklin & Ballan 2011), the empirical evidence submitted in this book underwent a thorough process of data refinement designed to rule out irregularities within the answers provided by the interviewees.

It is important to note that the primary data collected during these face-to-face interviews do not constitute the bulk of the empirical evidence utilised in support of this book's findings. The information shared by our interviewees rather integrates the extensive analysis of the literature available on the Russian COIN operations in Chechnya. Additionally, we complement the scrutiny of Russia's COIN experience by incorporating several supplementary case studies of brutalisation-centred COIN campaigns. The recourse to a wide array of different sources reinforces this book's findings and showcases the cross-case relevance of its arguments.

Contents

This book consists of an introduction, three core chapters, and a conclusion. This introductory chapter has outlined the book's overall aims and introduced the research design followed in the central chapters. We engaged in theoretical debates on the nature of COIN warfare to expose the blind spots affecting the literature and present the main topics covered in this book.

Chapter 2 conceptualises the fundamentals of population-centric COIN warfare. Drawing upon an extensive overview of the available literature, the chapter compares and contrasts the two principal strategic pathways available to counterinsurgents, known as the hearts and minds approach and the brutalisation paradigm. It explores the principal streams of research in the field of COIN studies, dives deep into the brutalisation paradigm, and identifies the main challenges to the study of brutalisation-centred COIN campaigns.

Chapter 3 outlines the conceptual framework utilised for the analysis of brutalisation-centred COIN campaigns. We submit an original visualisation of the pathway chosen by counterinsurgents following the brutalisation paradigm, which we define as the "shortcut approach," necessary to highlight where and how this paradigm diverges from the theory and practice of the hearts and minds approach. The chapter advances an in-depth analysis of the intelligence, information, military, political, and economic components of a population-centric COIN operation, explaining why, when, and how the brutalisation paradigm outperforms the hearts and minds approach. It then identifies what victory in COIN warfare looks like and defines the parameters necessary to assess the degree of effectiveness achieved by counterinsurgents waging population-centric COIN warfare.

Chapter 4 is dedicated to the analysis of the Russian COIN operations during the Second Russian-Chechen War of 1999–2009, which is further strengthened by the inclusion of several supporting case studies. It utilises the conceptual framework presented in the previous chapter to unravel and assess the brutalisation paradigm from an empirical perspective. Focusing on the major components of population-centric COIN campaigns, the chapter shows how counterinsurgents plan and carry out key operational tasks according to the logic of the brutalisation paradigm. The chapter assesses the results achieved by the Russians in Chechnya and problematises several narratives promoted by the advocates of the hearts and minds approach. The chapter concludes by advancing five key takeaways for Western practitioners. The book's main findings are summarised and tied together in Chapter 5, which offers additional avenues for further research into the brutalisation paradigm.

Notes

1 Interview with PG20201.
2 Interview with PG20202.
3 Interview with PG20204.

References

Avramov, Kiril. "Gifts from Grozny: The Export of the Russian COIN Model to Syria." *Small Wars Journal*, 2018. Available at: https://smallwarsjournal.com/jrnl/art/gifts-grozny-export-russian-coin-model-syria.

Blank, Stephen. "Russian Counterinsurgency in Perspective." In *Insurgencies and Counterinsurgencies: National Styles and Strategic Cultures*, edited by Heuser, Beatrice & Shamir, Eitan (Cambridge: Cambridge University Press, 2016). ISBN: 978-1-107-13504-8.

Boyatzis, Richard E. *Transforming Qualitative Information: Thematic Analysis and Code Development* (London: SAGE Publications, 1998). ISBN: 0.7619-0960-5.

Dowdall, Jonathan & Smith, M. L. R. "Counter-Insurgency in the Grey: The Ethical Challenge for Military Conduct." *Contemporary Security Policy* 31(1), 2010: 34–60. https://doi.org/10.1080/13523261003640850.

Dyer, Jr. Gibb & Wilkins, Alan L. "Better Stories, not Better Constructs, to Generate Better Theory: A Rejoinder to Eisenhardt." *Academy of Management Review* 16(3), 1991: 613–619.

Franklin, Cynthia & Ballan, Michelle. "Reliability and Validity in Qualitative Research." In *The Handbook of Social Work Research Methods*, edited by Thyer, Bruce A. (Thousand Oaks: SAGE Publications, 2011). https://doi.org/10.4135/9781412986182.

Glenn, Russell W. *Rethinking Western Approaches to Counterinsurgency: Lessons from Post-Colonial Conflict* (Oxon: Routledge, 2015). ISBN: 978-1-138-81933-7.

Haines, John R. "A Method to the Madness: The Logic of Russia's Syrian Counterinsurgency Strategy." *Foreign Policy Research Institute*, 5th January 2016. Available at: www.fpri.org/article/2016/01/method-madness-logic-russias-syrian-counterinsurgency-strategy/.

Human Rights Watch. *"Like Walking a Minefield": Vicious Crackdown on Critics in Russia's Chechen Republic*, August 2016. Available at: www.hrw.org/report/2016/08/31/walking-minefield/vicious-crackdown-critics-russias-chechen-republic.

ICRC (International Committee of the Red Cross). *Protocol Additional to the Geneva Conventions of 12 August 1949 and Relating to the Protection of Victims of International Armed Conflicts* (Protocol Additional I), 8th June 1977, art. 48. Available at: www.ohchr.org/en/professionalinterest/pages/protocolii.aspx.

Iliyasov, Marat. "Chechen Demographic Growth as a Reaction to Conflict: The Views of Chechens." *Europe-Asia Studies* 71(10), 2019: 1710. https://doi.org/10.1080/09668136.2019.1602593.

Jeffrey, James F. "Why Counterinsurgency Doesn't Work: The Problem Is the Strategy, Not the Execution." *Foreign Affairs* 94(2), 2015: 178–180.

Kahl, Colin H. "In the Crossfire or the Crosshairs? Norms, Civilian Casualties, and U.S. Conduct in Iraq." *International Security* 32(1), 2007: 7–46. Available at: www.jstor.org/stable/30129800.

Lavine, Daniel H. "Care and Counterinsurgency." *Journal of Military Ethics* 9(2), 2010: 139–159. https://doi.org/10.1080/15027570.2010.491331.

Maguire, Moira & Delahunt, Brid. "Doing a Thematic Analysis: A Practical, Step-by-Step Guide for Learning and Teaching Scholars." *AISHE-J* 8(3), 2017.

McLeod, Travers. *Rule of Law in War: International Law and United States Counterinsurgency in Iraq and Afghanistan* (Oxford: Oxford University Press, 2014). ISBN-13: 9780198716396.

Mockaitis, Thomas. "Winning Hearts and Minds in the 'War on Terrorism'." *Small Wars & Insurgencies* 14(1), 2003: 21–38. https://doi.org/10.1080/0959231042331300546.

Nagl, John A. "The Age of Unsatisfying Wars." *The New York Times*, 6th June 2012. Available at: www.nytimes.com/2012/06/07/opinion/the-age-of-unsatisfying-wars.html.

Nagl, John A. *Learning to Eat Soup with a Knife: Counterinsurgency Lessons from Malaya and Vietnam* (Chicago: The University of Chicago Press, 13th Edition, 2005). ISBN: 0-275-97695-5.

Paul Christopher & Clarke, Colin P. *Counterinsurgency Scorecard Update* (Santa Monica: RAND Corporation, 1st Edition, 2016). ISBN: 978-0-8330-9262-5.

Peters, Ralph. "The Heart-and-Minds Myth." *Armed Forces Journal*, 1st September 2006. Available at: http://armedforcesjournal.com/the-hearts-and-minds-myth/.

Putin, Vladimir. "Speech at Expanded Meeting of the State Council on Russia's Development Strategy Through to 2020." *The Kremlin*, 8th February 2008. Available at: http://en.kremlin.ru/events/president/transcripts/24825.

Ratelle, Jean-François. "Radical Islam and the Chechen War Spillover: A Political Ethnographic Reassessment of the Upsurge in the North Caucasus since 2009." PhD Thesis, University of Ottawa, 2013: 219–220.

Roxborough, Ian. "Counterinsurgency: The U.S. Military Should Have Learned a Lot About Fighting Rebels in Vietnam. So Why is Iraq Such a Disaster?" *Context* 6(2), 2007: 15–21.

Schaefer, Robert W. *The Insurgency in Chechnya and the North Caucasus: From Gazavat to Jihad* (Santa Barbara: Praeger Security International, 2010). ISBN: 978-0-313-38634-3.

Stack, Megan K. "Chechen Tiger Without a Chain." *Los Angeles Tiger*, 17th June 2008. Available at: www.latimes.com/archives/la-xpm-2008-jun-17-fg-kadyrov17-story.html.

Tacitus, *Dialogus Agricola Germania* (London: William Heinemann, 1914[98 AD]).

Tenzek, Kelly E. "Snowball Subject Recruitment." In *The SAGE Encyclopaedia of Communication Research Methods*, edited by Allen, Mike (Thousand Oaks: SAGE Publications, 2018). https://doi.org/10.4135/9781483381411.

U.S. Government. *Department of Defense Law of War Manual* (Washington, DC: Department of Defense, December 2016).

U.S. Government. *FM 3–24 MCWP 3–33.5: Insurgencies and Countering Insurgencies* (Washington, DC: Department of the Army, 2014).

U.S. Government. "Secretary Michael R. Pompeo At a Press Availability After the Afghanistan Signing Ceremony (February 29)." *U.S. Department of State*, 29th February 2020. Available at: www.state.gov/at-a-press-availability-after-the-afghanistan-signing-ceremony/.

Zhukov, Yuri M. "Counterinsurgency in a Non-Democratic State: The Russian Example." In *The Routledge Companion to Insurgency and Counterinsurgency*, edited by Rich, Paul B. & Duyvesteyn, Isabelle (London: Routledge, 2011). ISBN: 978-0-415-56733-6.

Zhukov, Yuri M. "Evaluating Success in Counterinsurgency, 1804–2000: Does Regime Type Matter?" Working Paper, 21st Convention of the International School on Disarmament and Research on Conflicts (ISODARCO), 2008. Available at: www.isodarco.it/courses/andalo08/paper/andalo08_Zhukov_paper.pdf.

2 Conceptualising population-centric counterinsurgency warfare

Introduction to insurgency and counterinsurgency warfare

In the field manuals issued to Western forces, the insurgency is defined as "an organised, violent and politically motivated activity conducted by non-state actors, sustained over a protracted period of time," and performed to "seize, nullify, or challenge the political control of a region" (Australian Gov. 2008: xx; U.S. Gov. 2018: GL-5). As the aim of rebel movements is to rule over a territory by overthrowing its established authority, insurgent warfare can be understood as "a process of alternative state-building" in which violence is utilised to catalyse the government's downfall and facilitate the establishment of a clandestine, insurgent-controlled political entity (Jones 2017: 8; Carter 2016: 136). This conceptualisation of insurgent warfare is echoed in the works of some of the most influential insurgent thinkers. Whilst Ernesto "El Che" Guevara contends that a protracted insurgency deprives the government of its credibility as a resilient, capable authority (Guevara 1964: 2; Payne 2011: 126), revolutionary leader Marighella argues that "the rebellion of the urban guerrilla . . . is the best way of ensuring public support for the cause" (Marighella 2011: 40). All successful insurgencies, such as the Vietcong insurgency in South Vietnam (1954–1976) and the Mujahideen insurgency in Afghanistan (1978–1992), realised that "people are the lifeblood of rebellion" and relied on the population's support to catalyse the process of regime change (Jardine 2012: 264). To confront, contain, and defeat insurgency, states perfected a vast selection of techniques known as counterinsurgency (COIN) strategies.

According to the U.S. Department of Defence (DoD), COIN warfare can be defined as "comprehensive civilian and military efforts designed to simultaneously defeat and contain insurgency" (U.S. Gov. 2019: 55). As the terminological broadness of this definition suggests, COIN constitutes an umbrella term for a wider spectrum of procedures that states utilise to

DOI: 10.4324/9781003188049-2

suppress and countervail rebellion. To coherently select from this list of techniques their preferred strategic options, counterinsurgents dispose of two main philosophies that provide guidance during the planification and execution of COIN operations, conventionally known as the enemy-centric and population-centric paradigms.

The leading thesis of the enemy-centric approach as described by population-centric sceptics, such as Luttwak (2007) and Gentile (2013a), is that COIN warfare abides by the same laws and principles modulating conventional warfare. Enemy-centric subscribers maintain that kinetically based strategies aimed at annihilating the enemy's fighting force can permanently incapacitate the insurgency and prevent local grassroots uprisings from turning into full-scale national rebellions (Plakoudas 2015: 132). Although military strategists frequently resume enemy-centric approaches when combat units are confronted by large enemy forces (Springer 2012), contemporary counterinsurgents abandoned enemy-centric notions in favour of more versatile population-centric blueprints of action.

The fundamental assumption behind the population-centric paradigm is the dismissal of the Clausewitzian axiom considering "the destruction of the enemy's physical force" as the linchpin for success (Clausewitz 1984: 71). Since the population constitutes the "centre of gravity" from which insurgents derive their moral and physical strength, population-centric promoters theorise that insurgent groups will remain undefeated, regardless of the number of casualties suffered, as long as they extract from the population enough resources and manpower to keep fighting against the incumbent (U.S. Gov. 2014: 7-6; Mansoor & Ulrich 2007: 21). Instead of advocating for killing the enemy at any cost, population-centric doctrines attest that re-establishing the government's exclusive control over the population asphyxiates the rebellion and severs the insurgents from their principal lifelines (Costa 2006: 7; Shy & Collier 1986: 820). Although these conceptual foundations guide the vast majority of contemporary COIN operations, the nature of population-centric COIN warfare cannot be fully grasped without considering the strategic approach chosen by a given counterinsurgent.

Hearts and minds versus brutalisation

To succeed in a population-centric COIN operation, military prowess does not suffice; a wider engagement is required. While the scholarship generally agrees that insurgency is best quelled when the rebels are deprived of their popular support, there is a sharp disagreement over which combination of military, political, and economic measures is the most effective.

Serving as a guiding paradigm for many contemporary COIN operations, the hearts and minds approach has experienced a soar in popularity ever since the U.S. military engagement in Afghanistan and Iraq. The origins of this term are attributed to Sir Gerald Templer, a British General tasked with defeating the Communist insurgency enraging in Malaya from 1948 to 1960, who stated: "the answer [to the uprising] lies not in pouring more troops into the jungle, but in the hearts and minds of the people" (Lapping 1985: 224). According to hearts and minds supporters, efforts aimed at gaining popular support should take priority over killing the enemy, as socio-economic paternalistic measures are believed to alleviate the population's discontent and help restore legitimacy for the incumbent (U.S. Gov. 2018: 1–3).

For a counterinsurgent, however, following the hearts and minds approach comes neither easy nor cheap. This is so for two principal reasons. The imperative of minimising the use of force and reducing collateral damage is the first challenge. As specified by Kilcullen, hearts and minds campaigns are unsuccessful if the counterinsurgent does not act in respect of the local people: "the well-being of non-combatant civilians [should be put] ahead of any other consideration" (Kilcullen 2010: 4). The extraordinary efforts required to rebuild a war-torn country anew is a second, equally challenging step necessary to gain the population's wholehearted support. As reported by a former British officer, the difficulties of promoting good governance are multifaceted:

> one must make a difference in the lives of as many people as possible as early as possible. This necessitates a significant effort to ensure fair treatment, the creation of jobs, improvements in education and medical services . . . providing a bearable standard of living, basic personal security, and some form of legitimate representative governance.
>
> (Garfield 2006: 16)

It follows that, in order to bear fruits, the hearts and minds approach requires the counterinsurgent to follow strict Rules of Engagement (ROE) designed to minimise the use of lethal force. As explained by Frank Kitson, another British General who advocated for the hearts and minds approach during the Malayan Emergency, failing to uphold the values of compassion, patience, and self-discipline brings about severe repercussions: "Those [officers] who are not capable of developing these characteristics are inclined to retreat into their military shells," and the population is likely to perceive their actions as "the ridiculous blunderings of a herd of elephants" (Kitson 1971: 200). In this struggle for acquiring the population's enthusiastic support, "shifting popular attitudes [and] sympathies . . . away from the insurgents and towards the government" constitutes the essence of a COIN endeavour focused on convincing people that "their best interests are served by COIN success . . . and that resisting is pointless" (Findley & Young 2007: 381; U.S. Gov. 2006: A-5).

If protecting the local population is the "cornerstone" of the hearts and minds approach (U.S. Gov. 2006: 1–23), the extensive use of coercion constitutes the centrepiece of the brutalisation paradigm. According to this variant of population-centric COIN warfare, winning over the population's favour is unnecessary, for one has to be feared—not loved—to be obeyed and respected. Considering that this approach assumes that the behavioural choices of individuals caught in the midst of a conflict are shaped by cost-benefit calculations (Waldron 2004; Kalyvas 2004), a counterinsurgent would find more success in punishing disloyalty over encouraging enthusiastic compliance. As sustained by Callwell in his early 20th century manual on "*Small Wars*," cowing people into submission accelerates the attainment of territorial control at the expense of the rebels' ability to garner popular support: "in the suppression of a rebellion the refractory subjects of the ruling power must all be chastised and subdued" (Callwell 1906: 147). Faced with unacceptable risks, the population will refrain from taking part in anti-incumbent activity, and the insurgents will no longer find supporters amongst a terrorised population. During the U.S. military expedition against the Tagalog insurgency in the Philippines (1899–1902), the U.S. servicemen targeted the local villagers to sever the insurgents from their popular mass base. As reported by General Franklin Bell, victimising civilians would have sent a clear message to the wider population:

> it is necessary to make the state of war as insupportable as possible, and there is no more efficacious way of accomplishing this than by keeping the minds of the people in such a state of anxiety . . . that living under such conditions will soon become unbearable. Little should be said . . . Let acts, not words, convey intentions.
>
> (U.S. Gov. 1902: 1628)

According to the brutalisation paradigm, striving to gain the population's endorsement is considered as a "military malpractice" that dissipates time and resources without inflicting visible damages on the insurgency's infrastructure (Luttwak 2007). As stated by a U.S. serviceman deployed in South Vietnam, kindness and empathy are redundant when facing a grass-root rebellion: "Grab 'em by the balls and their hearts and minds will follow" (Dixon 2009: 354).

The divide between concessive and coercive approaches has long defined the study of COIN warfare. Nevertheless, a minority of commentators have recently problematised this dichotomy and, on the basis of historical data, argued that the hearts and minds approach is only a façade that conceals the recourse to heavy-handed repression. According to this scholarship, considering the COIN experience of Western nations as a kinder, gentler warfare is a "strategic illusion" resting on false premises and superficial narratives of military effectiveness (Etzioni 2015). In their revisionist work carried out on the British COIN in Malaya, French (2011) and Miller (2012) argued that behind a rhetorical narrative centred on winning hearts and minds, this ideal type of COIN was characterised by mass arrests, torture, forcible population resettlements, and food denial operations. "It does not amuse me to punish innocent people," declared Templer in front of several local village leaders in 1952, "but many of you are not innocent. You have information which you are too cowardly to give" (Nagl 2005: 89). Following this revisionist agenda, Gurman sustained that the U.S. troops deployed in South Vietnam resorted to similar techniques, claiming that the hearts and minds propagandistic narrative concealed a "schizophrenic" COIN characterised by the "tendency for the carrot to become a stick" (Gurman 2013: 160). With episodes of brutality surfacing from some of the most notable hearts and minds COIN campaigns, many observers have concluded that history should not be "a bland cupboard from which to raid lessons learned which serve to confirm ideas already arrived at in the present" (Gumz 2009: 581).

Although these studies underscore that the hearts and minds approach rests on "mythologised" interpretations of the past, lessons learned from misdirected narratives have nevertheless created a contemporary repertoire of "gentler" COIN practices (Porch 2011). This recent, but enduring shift from coercive to concessive COIN measures is itself criticised by several researchers, who contend that the embrace of "distinctively liberal, humanistic values" is atrophying the liberal democracies' ability to wage warfare (Cohen 2010: 75). These claims find confirmation in the criticisms raised by veterans of the wars in Afghanistan and Iraq, where restrictive ROE are limiting the soldiers' ability to confront local guerrillas. While retired military officers from Iraq complain that strict ROE are preventing the combat troops from taking initiative against the insurgents (French 2015), the forces operating in Afghanistan are often ordered not to engage enemy forces because they are unable to demonstrate their "overt hostility" (Phillips 2016). Constrained by what Gentile (2010) defined as a strategic "straitjacket," counterinsurgents adhering to the hearts and minds approach discarded coercion as incompatible with the moral requirements of military operations that see the "human terrain" as the decisive operational battleground.

The challenges advanced against the historical accuracy of hearts and minds narratives stimulated a much-needed re-conceptualisation of todays' ways of warfare. Nevertheless, these arguments rest upon the acceptance that Western counterinsurgents rely on strategic formulas intrinsically incompatible with coercion-centred blueprints. "The story on which the current practice of COIN depends is not supported by evidence," writes Gentile (2013b: 34). But this does not mean that the hearts and minds approach has not made its way through the strategic thinking and everyday practice of modern-day counterinsurgents. By confirming that liberal democracies have drifted apart from coercion-centred measures, the existing literature indicates that the hearts and minds approach is deeply rooted in the theory and practice of contemporary COIN warfare. Hence, failing to distinguish between hearts and minds and brutalisation would provide an inaccurate picture of the strategies developed by Western states during the last two decades of uninterrupted COIN operations.

Brutalisation as an underresearched paradigm

Once relegated to the realm of military history, the study of COIN warfare experienced an intense intellectual renaissance following the Western democracies' involvement in the battlefields of Afghanistan and Iraq (Hussain 2010). Captivated by the opportunity of devising and refining the foundations of a 21st-century model of asymmetric warfare, the scholarship focused on delineating sets of "best COIN practices" customised for Western counterinsurgents, sentencing coercive approaches to receive scant attention and being profiled as checklists of "bad" COIN practices (Sepp 2005; Paul & Clarke 2016: 3). The ongoing debate on the hearts and minds approach sees the scholarship divided along two antagonistic camps.

Exponents of the "distinctively liberal" Western narrative consider the hearts and minds approach as inherently superior to its brutalisation-centred counterpart. In line with Nagl's (2007) conceptualisation of the "American way of COIN" as the most effective way to confront insurgency, Findley and Young assert that "the hearts and minds strategy consistently outperforms the attrition approach" (Findley & Young 2007: 379). This viewpoint echoes the work carried out by renowned hearts and minds advocates. In his influential monograph on *"Defeating Communist Insurgency,"* Sir Robert Thompson argues that victimising civilians only stiffens the population's resolve in supporting the insurgents: "a government which does not act in accordance with the law forfeits the right to be called a government and cannot then expect its people to obey the law" (Thompson 1966: 52). A successful counterinsurgent cannot be an oppressor, because a brutalised population will always choose the insurgents over a barbaric ruler: "resorting to firepower solutions *readily* becomes self-defeating" (Record 2007: 106, emphasis added). Shared by the majority of the scholarship expounding the logic of population-centric COIN warfare, this school of thought dismisses as detrimental any strategic approach conflicting with the ones endorsed by hearts and minds advocates. As emphasised by Mockaitis, battering the population into passive submission can only be a futile endeavour: "There is no substitute for an effective hearts-and-minds campaign" (Mockaitis 2003: 37).

On the other hand, critics of the hearts and minds approach maintain that Western counterinsurgents are suffering from a severe case of "strategic myopia" (Porch 2013: 309). In sustaining that self-inflicted restrictions on the use of force and the obsession for securing the population's approval are depleting the range of strategic options available to counterinsurgents, these authors warn practitioners against being bogged down in precepts with no historical record of success. In his "Requiem for American Counterinsurgency," Gentile criticises the deceitful guidelines issued to the troops in Afghanistan and Iraq, arguing that winning the population's hearts and minds should never be considered as a top-priority objective: "that idea is dead and should be banished into the graveyard of history's really bad ideas" (Gentile 2013a: 549). Gentile's position mirrors the one taken by Luttwak, who argues that COIN warfare cannot be performed without a good dose of coercion: "The capacity of American armed forces to inflict collective punishments does not extend much beyond curfews . . . inconvenient to be sure . . . but obviously insufficient to out-terrorize insurgents" (Luttwak 2007: 42). The dissatisfaction and frustration with the ways in which the U.S. engages in modern COIN warfare even prompted several critics to idolise certain aspects of the brutalisation paradigm. "You do not nation build, you don't try to hold ground," stated a retired U.S. Army lieutenant during an interview: "You go wherever in the world the terrorists are and you kill them, you do your best to exterminate them, and then you leave, and you leave behind smoking ruins and crying widows" (Fox News 2015). While such criticisms call for a re-evaluation of the hearts and minds approach, these commenters have equally failed to dive deeper into the brutalisation paradigm, plausibly because coercion-centred approaches are considered as inappropriate to impart useful lessons to Western counterinsurgents (Miroiu 2015: 179).

Although the vast majority of the scholarship has turned a blind eye to the study of brutalisation-centred COIN campaigns, a handful of researchers recently started expounding the logic behind coercion in COIN warfare. This emerging interest for the brutalisation paradigm was stimulated by provocative works conducted on the alleged (in)effectiveness of COIN operations performed by democratic regimes, of which Merom's monograph titled "*How Democracies Lose Small Wars*" constitutes the most vivid example. Tracing the causes of COIN failure in cultural and political processes occurring in the counterinsurgent's domestic political realm, Merom argues that morally driven regimes are ill-positioned to fight intrinsically vicious conflicts: "the most disturbing conclusion from our current moral vantage point is that brutality pays" (Merom 2003: 47).

Yet, given the lack of empirical enquiry into this topic, Merom's conclusions can only be considered as tentative. In an effort to shed light on a research area that has "largely eluded systematic empirical investigation," Zhukov (2007, 2011: 287) scrutinised the Soviet Union's COIN experience and produced the first publications in which brutalisation features as a standalone conceptual paradigm. If Zhukov pioneered this stream of research by introducing a baseline conceptualisation of the brutalisation paradigm, his research only marginally discussed non-kinetic aspects, with the consequence that several of its components were left unexplored. Substantial progress occurred rather recently, when Byman (2016) and Ucko (2016) submitted two studies that offer critical insight into the brutalisation paradigm. In juxtaposing the repertoire of brutalisation-centred measures with the theories of COIN warfare as formulated by Western strategists, Byman and Ucko expanded Zhukov's work, demonstrating that coercion-centred endeavours cannot be reduced to a mere display of sheer force and wanton cruelty.

Despite constituting a major breakthrough for the study of the brutalisation paradigm, neither of the two studies attempted to trace the conceptual roots on which any brutalisation-centred COIN campaign rests upon. While the two authors extensively covered the most prominent aspects of the brutalisation paradigm, the lack of a solid analytical framework left undisclosed other, more subtle but equally important features. For Byman, the brutalisation paradigm can be distilled down to the skilful use of repression, aggressive intelligence techniques, and massive propaganda campaigns followed by population transfers and ethnic cleansings when less aggressive measures are deemed insufficient. According to the author, political and economic measures find no substantial application in a brutalisation-centred COIN campaign: "[these counterinsurgents] neglect administrative capacity building and economic development—or, when they do so, it is overshadowed by the constant threat of violence" (Byman 2016: 85). While it is true that violence constitutes the centrepiece of the brutalisation paradigm, dismissing important components of a COIN campaign as a "flawed enterprise at best" does little to advance the collective understanding of unorthodox variants of COIN warfare (Ibid.). A similar narrative runs through Ucko's account of the brutalisation paradigm. Although the author acknowledges that building political legitimacy and promoting economic development are endeavours often undertaken in brutalisation-centred COIN campaigns, the centrality of mass violence and repression incentivised the author to downplay the relevance of non-kinetic components: "their impact has often been minimal or meaningless" (Ucko 2016: 50). Given that Byman and Ucko's publications constitute a point of reference for any researcher approaching this relatively new field of studies, filling the gaps left unaddressed by the two assumes a role of primary importance. Building on Byman and Ucko's work, the present book enriches this underdeveloped research area and sheds light on long-ignored aspects of the brutalisation paradigm.

References

Australian Government. *Land Warfare Doctrine LWD 3–0–1. Counterinsurgency* (Australian Army, 2008).

Byman, Daniel. "'Death Solves All Problems': The Authoritarian Model of Counterinsurgency." *Journal of Strategic Studies* 39(1), 2016: 62–93. https://doi.org/10.1080/01402390.2015.1068166.

Callwell, C. E. *Small Wars: Their Principles and Practice* (London: Harrison and Sons, 3rd Edition, 1906).

Carter, David B. "Provocation and the Strategy of Terrorist and Guerrilla Attacks." *International Organisation* 79, Winter 2016: 133–173. https://doi.org/10.1017/S0020818315000351.

Clausewitz, Von Carl. *On War* (London: Everyman's Library, 1984). ISBN: 0-691-05657-9.

Cohen, Michael A. "The Myth of a Kinder, Gentler War." *World Policy Institute*, Spring 2010: 75–86.

Costa, Christopher P. "Phoenix Rises Again: HUMINT Lessons for Counterinsurgency Operations." *Joint Military Operations Department, Naval War College*, 16th May 2006.

Dixon, Paul. "'Hearts and Minds'? British Counter-Insurgency from Malaya to Iraq." *Journal of Strategic Studies* 32(3), 2009: 353–381. https://doi.org/10.1080/01402390902928172.

Etzioni, Amitai. "COIN: A Study of Strategic Illusion." *Small Wars & Insurgencies* 26(3), 2015: 345–376. https://doi.org/10.1080/09592318.2014.982882.

Findley, Michael G. & Young, Joseph K. "Fighting Fire with Fire? How (Not) to Neutralize an Insurgency." *Civil Wars* 9(4), 2007: 378–401. https://doi.org/10.1080/13698240701699482.

Fox News. *Peter's Plan to Fight Terror: 'Leave Behind Smoking Ruins and Crying Widows'*, 9th January 2015. Available at: https://insider.foxnews.com/2015/01/09/lt-col-peters-plan-fight-terror-leave-behind-smoking-ruins-crying-widows.

Frank, Kitson. *Low Intensity Operations: Subversion, Insurgency, Peace-Keeping* (London: Faber and Faber, 1971).

French, David. *The British Way in Counter-Insurgency 1945–1967* (Oxford: Oxford University Press, 1st Edition, 2011). https://doi.org/10.1093/acprof:oso/9780199587964.001.0001.

French, David. "How Our Overly Restrictive Rules of Engagement Keep Us from Winning Wars." *National Review*, 21st December 2015. Available at: www.nationalreview.com/2015/12/rules-engagement-need-reform/.

Garfield, Andrew. *Succeeding in Phase IV: British Perspectives on the U.S. Effort to Stabilize and Reconstruct Iraq* (Philadelphia: Foreign Policy Research Institute, 2006).

Gentile, Gian P. "Freeing the Army from The Counterinsurgency Straitjacket." *Joint Force Quarterly* 58, 2010: 121–122.

Gentile, Gian P. "A Requiem for American Counterinsurgency." *Orbis*, Autumn 2013a: 549–558. https://doi.org/10.1016/j.orbis.2013.08.003.

Gentile, Gian P. *Wrong Turn: America's Deadly Embrace of Counterinsurgency* (New York: The New Press, 1st Edition, 2013b). ISBN: 978-1-59558-874-6.

Guevara, Che Ernesto. *Guerrilla Warfare* (Beijing: Foreign Languages Press, 1964).

Gumz, Jonathan E. "Reframing the Historical Problematic of Insurgency: How the Professional Military Literature Created a New History and Missed the Past." *Journal of Strategic Studies* 32(4), 2009: 553–588. https://doi.org/10.1080/01402390902986972.

Gurman, Hannah. *Hearts and Minds: A People's History of Counterinsurgency* (New York: The New Press, 1st Edition, 2013). ISBN: 978-1-59558-843-2.

Hussain, Nasser. "Counterinsurgency's Comeback: Can a Colonialist Strategy Be Reinvented?" *Boston Review* 35(1), January–February 2010: 23–26.

Jardine, Eric. "Population-Centric Counterinsurgency and the Movement of Peoples." *Small Wars and Insurgencies* 23(2), 2012: 264–294. https://doi.org/10.1080/09592318.2012.642220.

Jones, Seth G. *Waging Insurgent Warfare: Lessons from the Viet Cong to the Islamic State* (London: Oxford University Press, 2017). ISBN: 9780190600860.

Kalyvas, Stathis N. "The Paradox of Terrorism in Civil War." *The Journal of Ethics* 8, 2004: 97–138.

Kilcullen, David. *Counterinsurgency* (London: C. Hurst & Company, 2010). ISBN: 978-0-19-973748-2; 978-0-19-973749-9.

Lapping, Brian. *End of Empire* (London: Granada, 1985). ISBN: 0312250711.

Luttwak, Edward N. "Dead End: Counterinsurgency Warfare as a Military Malpractice." *Harper's Magazine*, February 2007: 33–42.

Mansoor, Peter R. & Ulrich, Mark S. "Linking Doctrine to Action: A New COIN Center-of-Gravity Analysis." *Military Review*, September–October 2007.

Marighella, Carlos. "Appendix-Mini-Manual of the Urban Guerrilla." In *No Exit: North Korea, Nuclear Weapons and International Security*, edited by Pollack, Jonathan D. (New York: Routledge, 2011). ISBN: 9780415670838.

Merom, Gil. *How Democracies Lose Small Wars: State, Society, and the Failures of France in Algeria, Israel in Lebanon, and the United States in Vietnam* (Cambridge: Cambridge University Press, 2003). ISBN: 0-52-1-80403-5.

Miller, Sergio. "Malaya: The Myth of Hearts and Minds." *Small Wars Journal*, 2012. Available at: https://smallwarsjournal.com/jrnl/art/malaya-the-myth-of-hearts-and-minds.

Miroiu, Andrei. "Deportations and Counterinsurgency: A Comparison of Malaya, Algeria and Romania." *Romanian Political Science Review* 15(2), 2015: 177–194.

Mockaitis, Thomas. "Winning Hearts and Minds in the 'War on Terrorism'." *Small Wars & Insurgencies* 14(1), 2003: 21–38. https://doi.org/10.1080/0959231041233130054.

Nagl, John A. "An American View of Twenty-First Century Counter-Insurgency." *The RUSI Journal* 152(4), 2007: 12–16.

Nagl, John A. *Learning to Eat Soup with a Knife: Counterinsurgency Lessons from Malaya and Vietnam* (Chicago: The University of Chicago Press, 13th Edition, 2005). ISBN: 0-275-97695-5.

Paul, Christopher & Clarke, Colin P. *Counterinsurgency Scorecard Update* (Santa Monica: RAND Corporation, 1st Edition, 2016). ISBN: 978-0-8330-9262-5.

Payne, Kenneth. "Building the Base: Al Qaeda's Focoist Strategy." *Studies in Conflict and Terrorism* 34(2), 2011: 124–143. https://doi.org/10.1080/10576 10X.2011.538832.

Phillips, Michael M. "Afghan War Rules Leave U.S. Troops Wondering When It's Ok to Shoot." *The Wall Street Journal*, 20th June 2016. Available at: www.wsj.com/articles/afghan-war-rules-leave-u-s-troops-wondering-when-its-ok-to-shoot-1466435019.

Plakoudas, Spyridon. "Strategy in Counterinsurgency: A Distilled Approach." *Studies in Conflict & Terrorism* 38(2), 2015: 132–145. https://doi.org/10.1080/10576 10X.2014.977000.

Porch, Douglas. *Counterinsurgency: Exposing the Myths of the New Way of War* (Cambridge: Cambridge University Press, 2013). ISBN: 978-1-107-69984-7.

Porch, Douglas. "The Dangerous Myths and Dubious Promise of COIN." *Small Wars & Insurgencies* 22(2), 2011: 239–257. https://doi.org/10.1080/09592318. 2011.574490.

Record, Jeffrey. *Beating Goliath: Why Insurgencies Win* (Washington: Potomac Books, 2007). ISBN: 978-1-59797-090-7.

Sepp, Kalev I. "Best Practices in Counterinsurgency," *Military Review*, May–June 2005: 8–12.

Shy, John & Collier, Thomas W. "Revolutionary War." In *Makers of Modern Strategy*, edited by Paret, Peter (Oxford: Oxford University Press, 1986). ISBN: 0-691-09235-4.

Springer, Nathan R. *Stabilizing the Debate between Population and Enemy-Centric Counterinsurgency: Success Demands a Balanced Approach* (Fort Leavenworth: Combat Studies Institute Press, 2012).

Thompson, Robert. *Defeating Communist Insurgency: The Lessons of Malaya and Vietnam* (New York: Praeger, 1966).

U.S. Government. *Dictionary of Military and Associated Terms* (Washington, DC: Department of Defence, 2019).

U.S. Government. *FM 3–24 MCWP 3–33.5 Counterinsurgency* (Washington, DC: Department of the Army, December 2006).

U.S. Government. *FM 3–24 MCWP 3–33.5: Insurgencies and Countering Insurgencies* (Washington, DC: Department of the Army, 2014).

U.S. Government. *Hearing before the Committee on the Philippines of the United States Senate: Affairs in the Philippine Islands* (Washington, DC: Government Printing Office, 1902).

U.S. Government. *Joint Publication 3–24: Counterinsurgency* (Washington, DC: Department of the Army, 2018).

Ucko, David H. "'The People are Revolting': An Anatomy of Authoritarian Counterinsurgency." *Journal of Strategic Studies* 39(1), 2016: 29–61. https://doi.org/ 10.1080/01402390.2015.1094390.

Waldron, Jeremy. "Terrorism and the Uses of Terror." *The Journal of Ethics* 8, 2004: 5–35. Available at: www.jstor.org/stable/25115779?seq=1#metadata_info_tab_contents.

Zhukov, Yuri M. "Counterinsurgency in a Non-Democratic State: the Russian Example." In *The Routledge Companion to Insurgency and Counterinsurgency*, edited by Rich, Paul B. & Duyvesteyn, Isabelle (London: Routledge, 2011). ISBN: 978-0-415-56733-6.

Zhukov, Yuri M. "Examining the Authoritarian Model of Counter-Insurgency: The Soviet Campaign Against the Ukrainian Insurgent Army." *Small Wars & Insurgencies* 18(3), 2007: 439–466. https://doi.org/10.1080/09592310701674416.

3 Theory and practice of brutalisation in counterinsurgency warfare

Introduction to the brutalisation shortcut

According to hearts and minds proponents, only second-class counterinsurgents would resort to brutalisation-centred measures. Yet, quantitative studies carried out on the duration and effectiveness of COIN campaigns found no evidence in support of the claim that brutalisation-centred COIN campaigns are flawed by design. Utilising a dataset of 168 asymmetric wars fought between 1945 and 2005, Lyall found that "the degree of a military's mechanization, its status as an external occupier, and the level of material support for insurgents" provide a "better theoretical bet" than other variables for explaining success in COIN warfare (Lyall 2010: 189). Elaborating on Lyall's dataset, Zhukov demonstrated that heavy-handed measures are far from being necessarily correlated with defeat. Of the 79 most noticeable episodes of forcible population resettlement recorded since 1917, only 11 "either failed to suppress an insurgency or could not prevent its recurrence" (Zhukov 2011: 293). These findings reason with a recent study by Asal and colleagues, in which the authors show that counterinsurgents refusing to be "as repressive and ruthless as might be necessary to win" are 40.3 percent less likely to prevail against grassroots insurgencies (Asal et al. 2017: 926). While statistical differences might lose their explanatory power when controlled for macro-historical, political, and environmental trends, these figures indicate that brutalisation-centred measures might be *at least* as effective as hearts and minds techniques (Getmansky 2013: 726; Johnston & Urlacher 2021).

These findings are puzzling because, according to Western COIN manuals, the unlawful use of excessive force, the deliberate targeting of innocent civilians and the uninterest in building bottom-up legitimacy constitute the perfect recipe for mission failure (U.S. Gov. 2006: 1–29). The incongruences between hearts and minds precepts and the evidence emerging from brutalisation-centred COIN campaigns suggest that the latter might be an equally valid pathway to success. Developed to bypass the operational and moral constraints regulating

DOI: 10.4324/9781003188049-3

the conduct of hearts and minds COIN campaigns, this "unorthodox" variant allows incumbents to suppress the insurgency in spite of their disregard for the civilian populations' well-being. Put otherwise, hearts and minds blueprints advance procedural elements that can be circumvented, altered, or removed with little to no repercussions for the overall COIN effort.

As this variant of population-centric COIN is built upon the premise that hearts and minds guidelines are either excessively restrictive or unnecessarily intricate, the brutalisation paradigm can be visualised as following a "shortcut" pattern in its execution. In proposing this visualisation, this book re-elaborates the ideas introduced by David Galula (2006[1964]) in *Counterinsurgency Warfare: Theory and Practice*, by many considered as the volume that first theorised population-centric COIN warfare (Cohen 2012). In his book, Galula identifies two principal strategic approaches followed by insurgent groups, defined as "orthodox" and "shortcut" variants, with the shortcut configuration sidestepping several orthodox practices throughout its execution (Galula 2006[1964]: 30–41). We transplant Galula's concept to the study of COIN operations as a means to illustrate where, why, and how brutalisation-centred measures diverge from the hearts and minds approach.

This book explores the logic of the brutalisation paradigm with the assistance of the theoretical framework designed by former U.S. Counterinsurgency advisor David Kilcullen (2006) in the article "Three Pillars of Counterinsurgency." Kilcullen identifies three pillars that, alongside intelligence and information operations, constitute the lines of action for achieving success in COIN warfare, conceptualised as the marginalisation of the insurgents and the attainment of uncontested control over the population. Although this framework was designed to accommodate the hearts and minds approach, its main precepts can be equally applied to brutalisation-centred COIN campaigns, as an additional proof of the fact that the two find a common theoretical ground. Just as the hearts and minds approach relies on the guidelines identified by Kilcullen, so does the brutalisation paradigm find an anchor in the pillars of population-centric COIN warfare. To enhance this chapter's intelligibility, each section builds upon the dichotomy between hearts and minds and brutalisation as earlier described in this book. The rationale supporting this choice is twofold. First, delineating the brutalisation paradigm without comparing and contrasting it with its counterpart would fail to highlight the principal differences between the two. Second, selecting a comparative approach helps underscoring that the brutalisation paradigm finds roots in the same theoretical foundations of the hearts and minds approach. By structuring the analysis around a clear-cut differentiation between the two paradigms, this chapter provides additional confirmation to Byman's intuition that Western states articulated "*a* model for counterinsurgency, but they should not believe it is the *only* model for success" (Byman 2016: 87).

Support base: intelligence penetration and information operations

At its essence, COIN warfare is an intelligence-driven endeavour in which the information submitted by indigenous informants constitutes a war-winning asset (U.S. Gov. 2019a: 115). Given that guerrilla warfare is waged "amongst the people," counterinsurgents can discriminate rebel combatants from innocent civilians only when local human sources (HUMINT) submit information on the rebels' identities, actions, and whereabouts (Smith 2007; Jackson 2007: 74). Although every counterinsurgent considers the collection of reliable intelligence as its "absolute highest priority" (Smith 2006), the methods chosen for collecting information diverge according to whether the population is seen as a partner to coax or as a target to strike.

According to hearts and minds proponents, high-quality HUMINT is retrieved only when the counterinsurgent protects its informants from the rebels' violent retaliation. This is so because people collaborate with the incumbent only when they have a reason to believe that no repercussion will follow for their actions. This principle refers to Galula's understanding of intelligence operations as successful only when the counterinsurgent protects the residents of an area affected by insurgent violence, as "the population will not talk unless it feels safe, and it does not feel safe until the insurgents' power has been broken" (Galula 2006[1964]: 50). Nevertheless, achieving this objective requires the consolidation of trust networks with the population and the continuous provision of physical security to local communities, both exhausting activities that produce actionable intelligence predominantly on a long-term basis (González 2018). As argued by the U.S. Major Jim Gant, one cannot expect to obtain high-quality intelligence upfront, because only "*after* a relationship has been built with the tribes, we will be able to gather relevant and actionable intelligence" (Gant 2009: 38, emphasis added). While building long-standing relationships with the population facilitates the collection of HUMINT, placing excessive trust in the local population's willingness to collaborate often does more harm than good. During the height of the combat operations in Baghdad, U.S. Army Colonel Peter Mansoor reported that voluntary submission of insurgent-related intelligence was often either deceitful or irrelevant:

> [the locals] offered lots of information but sorting truth from fiction and figuring out their motivations were extremely difficult . . . I often joked that in Baghdad we finally discovered something that travels faster than the speed of light—rumors, dubbed RUMINT [rumor intelligence].
> (Mansoor 2008: 47)

In Iraq, the frustration for the ineffectiveness of the available HUMINT even led the U.S. forces to "take the gloves off" and opt for more aggressive information-gathering measures. "You have to treat these detainees like dogs," allegedly stated Geoffrey Miller, the U.S. Army General managing the prison of Abu Ghraib, where suspected Iraqi insurgents were detained and subjected to harassment and torture for intelligence-gathering purposes (McKelvey 2007: 12).

While counterinsurgents following the hearts and minds approach strive to forge mutual interdependencies with local informants, those who opt for the brutalisation paradigm rely on their coercive potential to overcome conditions of "information starvation" and gather the intelligence required for the COIN effort (Lyall & Wilson 2009: 75). As observed by Byman, popular goodwill is only one of the possible ways to generate high-quality HUMINT: "blackmail, vendettas, bribes, and other less savoury forms" of deep intelligence penetration can produce equally abundant streams of high-quality information (Byman 2016: 76). In his account of the role of secret services in COIN operations, Ucko further confirmed that counterinsurgents frequently set up "all-seeing, all-hearing" police apparatuses to swiftly detect and wipe out insurgent activity (Ucko 2016: 46). These claims find rich empirical evidence in the literature exploring the *modus operandi* of intelligence agencies operating according to the brutalisation paradigm. To counter the insurgencies that emerged across the lands annexed after the Molotov-Ribbentrop Pact of 1939, Moscow established an enormous network of locally recruited agents to pinpoint and neutralise subversive elements within the society. As remarked by a Ukrainian insurgent, the Soviets set up such a dense network of collaborators that passing undetected was impossible:

> The informer network works meticulously. It is impossible to move across the area without being spotted . . . Informers are trained as well as circus dogs . . . No hideout unknown to [the Soviet Ministry of State Security] MVD-MGB exists in the village.
>
> (Statiev 2010: 238)

The Soviet Union's experience in the Western Borderlands was replicated by the Romanian Communist Party during the counterguerrilla operations performed against the partisans in the 1940s. Rather than blindly deploying its soldiers in high-risk combat operations, the Romanian government focused on creating a tight network of local informants capable of providing personalised information on the insurgents' identities and whereabouts. By recruiting informants "among the relatives of the partisans" and by striving "to get the support of poor peasants for their actions, to recruit shepherds, forest workers, and local guards," the Communist regime obtained the information necessary to locate and neutralise insurgent fighters and civilian supporters (Miroiu 2010: 678). More recent instances of similar intelligence procedures emerge from across numerous theatres of COIN warfare. To counter the Islamist insurgency rooted in the province of Sinai, the Egyptian government systematically tortures alleged civilian collaborators to obtain the names of people implicated in the insurgency. As reported by a former detainee, the torturers do not stop until the prisoners submit actionable information: "They electrocuted us in the genitals for hours before asking any questions. Then the torture continues during and after the interrogations" (Ashour 2019: 543). In Syria, the regime's secret services have made of torture their operational trademark. As reported by a captured manifestant, the officers would recur to a mix of psychological and physical abuses to break the detainees' resolve and make them confess. "I want you to turn this guy into art. I want to see colours on his skin," overheard a detainee from the room in which he was held captive (Abouzeid 2018: 285). By weaponising fear against the population, counterinsurgents can bypass the operational restrictions imposed by the hearts and minds approach and still receive large volumes of actionable HUMINT.

In addition to coercive intelligence techniques, a counterinsurgent following the brutalisation paradigm avails of Information Operations (IOs) to seize control over the population and deprive the rebels of their popular support. Defined by the DoD as the integrated deployment of "information-related capabilities . . . to influence, disrupt, corrupt, or usurp the decision-making of adversaries," IO is particularly suited to sabotage the insurgents' propaganda and promote pro-government political narratives (U.S. Gov. 2014b: Glossary-3). As with intelligence penetration techniques, the strategic approach chosen by a given counterinsurgent markedly affects the planification and execution of IO.

In hearts-and-minds COIN campaigns, IO constitutes a "soft power" instrument that counterinsurgents utilise to promote pro-incumbent political narratives "through attraction and persuasion rather than [with] threats of coercion" (Joseph 2016: 2; Nye 2017: 1). As specified in the COIN guidelines written by General McChrystal and General Petraeus for the Coalition forces deployed in Afghanistan, turning the population's perceptions "from fear and uncertainty to trust and confidence" requires the counterinsurgent to "stay true to his values" of compassion and empathy when performing COIN operations (NATO 2010a: 4; NATO 2010b: 3). As troops are given orders to refrain from engaging in unlawful or morally unacceptable activities, this approach aims at turning soldiers into culturally aware "good guests" keen at respecting local habits and customs. We define this strategy based on creating a synergy between words and actions as the "qualitative approach" to IO. As underscored by Armistead in his analysis of the IO performed by the U.S. forces in Yugoslavia, counterinsurgents put their credibility "on the line" when disseminating messages centred on values of compassion, peace, and cooperation, which would sound meaningless to the population if the troops fail to abide by strict behavioural and ethical precepts (Armistead 2004: 101). These conceptual guidelines have found their application in the strategic approach followed by Western forces deployed overseas. Reflecting on his experience as a leader of a Provincial Reconstruction Team (PRT) in Afghanistan, U.S. Commander Larry LeGree demonstrates that hearts and minds precepts are deeply rooted in the U.S. armed forces' approach to IO:

> The counterinsurgent must transmit the facts to the people as soon as possible as well as take the time to get the facts straight. We do more harm when we damage our own credibility than we do through any single lethal mistake or accident . . . Credibility is too important. Rigor in truth wins.
>
> (LeGree 2010: 24–25)

This was confirmed by another U.S. officer serving in an IO team in Afghanistan, who stated: "Adhering to cultural norms of politeness is a crucial step toward productive communication" (Munoz 2012: 107).

Although the mainstream scholarship considers IO narratives informed by hearts and minds precepts as excellent force multipliers (U.S. Gov. 2009a: C-7; Roca 2008: 34), qualitative approaches can give rise to significant backlashes if the counterinsurgent fails to honour the promises made to the population (Collings & Rohozinski 2005: ix). This potential drawback does not affect the IO performed according to the brutalisation paradigm, which departs from the interpretation of IO as a painless war instrument to embrace an approach defined by Van Herpen as "hard power in a velvet glove" (Van Herpen 2016: 40). Interpreting IO as an extension of military might, counterinsurgents following the brutalisation paradigm expose the population to incessant propaganda campaigns designed to suffocate the rebels' political machine and depict the insurgents as the sole actor responsible for the country's descent into lawlessness (Copeland & Potter 2008). During Japan's military occupation of Manchuria (1931–1940), the Japanese saturated the area of operations with brainwashing material meant to persuade the local communities to distance themselves from the insurgents' cause: "To Japanese officialdom in the 1930s, propaganda meant the cultivation of cultural values and attitudes that would be held so deeply they would appear innate and not imposed" (Kusher 2006: 25). Ever since the 1930s, in the Chinese province of Xinyang, the Chinese government has been waging an aggressive IO campaign to force the local ethnic group, known as Uyghur, to welcome the mainstream Chinese culture as their own cultural identity. Rounded up in re-education centres, ethnic Uyghurs are subjected to intensive "de-extremification" efforts achieved through "ethnic unity education, psychological counselling, lectures on government policy" and many other activities meant to detach the population from insurgent-sponsored political narratives (Zenz 2019: 114). Through a massive use of brainwashing propaganda, the counterinsurgent asphyxiates the enemy's mass-media infrastructure and saturates the area of operations with pro-government indoctrination material. Distinctive of the shortcut pattern, this "quantitative approach" to IO circumvents the complexity of the hearts and minds approach to fully exploit the counterinsurgent's superior propaganda machine. Applied in concert with well-coordinated military, political, and economic measures, aggressive intelligence techniques and invasive IO provide a powerful support base on which the brutalisation paradigm unfolds.

First pillar: use of force

Of the many axioms that the *FM 3–24* uses to describe the fundamentals of population-centric COIN warfare, prominent is the one regulating the use of lethal force against the rebels and their subterranean network of civilian supporters. Incorporating the British COIN principle of "minimum necessary force" (U.K. Gov. 2009: 3–13), the latest U.S. COIN doctrine contends that "the more force is used, the less effective it is" (U.S. Gov. 2014a: 7-2). In the words of U.S. Brigadier Michael Addison, "a sledgehammer should not be used to crack a nut because the sledgehammer might be better employed elsewhere. The doctrine of 'minimum force' forbids the use of a sledgehammer to protect the nut inside the shell" (quoted in Nagl 2005: 30). The strategic significance of self-imposed restrictions on the use of firepower has been championed by voluminous research conducted on the hearts and minds approach.

A dominant view within the hearts and minds camp is that indiscriminate violence—best defined by Lyall as "the collective targeting of a population without credible effort to distinguish between combatants and civilians" (Lyall 2009: 358)—is "at best ineffective and at worst counterproductive" (Kalyvas 2004: 112). Because strategies of blind civilian victimisation target individuals irrespectively of their non-collaboration with the opponent, scholars sustain that indiscriminate violence can engender an "inflammatory" effect on rebel activity (Zhukov 2014). As individuals fearing to fall victims to the government's randomised attacks do not gain benefits from complying with the incumbent's requests, a regime of wanton state terror incentivises people to increase their chances of survival by seeking the protection of the opponent's camp (Kalyvas 1999: 251). Given that non-combatants are often motivated to join insurgent groups out of the desire to avenge their relatives unjustly killed by the security forces, the literature expounding the causes of pro-insurgent mobilisation maintains that strategies of blind violence often lower the "societal costs of continued resistance" (Jones 2017: 47). As demonstrated by Kilcullen, neutral bystanders can turn into resolute guerrillas overnight and fight alongside hardcore rebels to seek vengeance for the incumbent's wrongdoings (Kilcullen 2009: 38). Although these considerations brought many experts to define civilian victimisation as categorically ineffective (Valentino et al. 2004; Hultquist 2017; Pechenkina et al. 2019), the same scholarship has also failed to consider *how* indiscriminate force is deployed in contexts of COIN warfare, with the consequence that coercion-centred approaches are often misinterpreted as raw displays of blind savagery.

Yet, in-depth research conducted on the logic of brutalisation-centred measures shows that strategies of civilian victimisation can reach remarkable degrees of rational sophistication. As remarked by Downes, the use of random violence against civilians is often the prelude—not the mainstay—of COIN operations. Utilised as a strategy of "early resort" against populations openly hostile to the incumbent, the extensive use of randomised violence in the early stages of the COIN effort allows counterinsurgents to establish a footprint on enemy territory when "there is little or no possibility" of gaining support from the local population (Downes 2006: 168). Since at the very start of a COIN campaign, information is scarce and therefore insufficient to carry out pinpointed strikes, harming the population—a readily accessible target—constitutes a strategy from which the counterinsurgent obtains immediate benefits in return. As argued by a 20th-century strategist Giulio Douhet, "spreading terror and havoc" across the country demoralises the enemy and shatters the population's spirit of resistance (Douhet 1983[1942]: 27). Later, we identify the ways in which counterinsurgents effectively victimise civilian populations to seize territorial control and sever the insurgents from their popular mass base.

Indiscriminate violence

In a fervid assault against the foundational doctrines of the hearts and minds approach, military historian Ralph Peters accuses the *FM 3–24* drafters of having either misunderstood or fictionalised the past 3,000 years of military history: "The teething-ring nonsense that insurgencies don't have military solutions defies history—it's campus and think-tank nonsense" (Peters 2007: 95). Despite hearts and minds advocates arguing that victimising civilian populations only leads to defeat, empirical evidence confirms indiscriminate violence as an effective COIN strategy of the early resort.

During their campaigns of territorial expansion, the Romans systematically brutalised indigenous populations to force local rebels and their fellow tribesmen into immediate capitulation. In his account of the Gallic Wars (58–50 BC), Caesar explains that devastating the territories inhabited by rebel tribes had a paralysing effect on the insurgency:

> Caesar tarried for a few days in their territory, until he had burnt all the villages and buildings, and cut down the corn-crops . . . he had accomplished all the objectives for which he had determined to lead his army across the Rhine—to strike terror into the Germans, to take vengeance on the [rebel tribe] Sugambri, to deliver the [loyal tribe] Ubii from a state of blockade.
>
> (Caesar 1919: 203–205)

Waging a campaign of indiscriminate warfare has often been the choice of preference for counterinsurgents whose interest is to suppress an insurgency regardless of the toll paid in innocent lives. During the U.S. government's repression of Native American tribes, little mercy was given to indigenous non-combatants caught in the midst of a military operation. As explained by General William Sherman, the end justified the means in the fight against the Indians: "We must act with vindictive earnestness against the Sioux, even to their extermination, men, women and children . . . during an assault, the soldiers cannot pause to distinguish between male and female, or even discriminate as to age" (Marszalek 2007 [1993]: 379). More recently, scorched earth strategies were followed by pro-government forces during the Guatemalan Civil War (1960–1996). To eradicate the insurgency's popular mass base, the Guatemalan government aimed at wiping out "entire sectors of the Maya population" that supported the insurgency (Jonas 2013: 359). As accounted by a local Mayan villager, the government forces stopped at nothing to crush the rebellion:

> the soldiers started grabbing people and torturing them, claiming that they were supporting the guerrillas. After torturing people, the soldiers would leave their bodies in public so everyone could see what would happen if you [were] associated with the guerrillas.
>
> (Marcucci 2017: 24)

Although appalling, scorched earth measures show an impressive trail of successes in past and present COIN operations. As explained by Grossman, episodes of methodical human savagery leave people in such a state of despair that resistance starts being pointless: "One of the most obvious and blatant benefits of atrocity is that it quite simply scares the hell out of people. The raw horror and savagery . . . cause people to flee, hide, and defend themselves feebly" (Grossman 1996: 207). While hearts and minds proponents are eager to state that "the more force is used, the less effective it is," cases of brutalisation-centred COIN campaigns demonstrate that counterinsurgents dispose of a callous, yet effective shortcut for securing an early success against grassroots rebellions (U.S. Gov. 2014a: 7-1).

Selective violence

According to the hearts and minds approach, counterinsurgents should place the utmost care in limiting collateral casualties amongst civilians when performing kinetic operations. As stated by a senior U.S. official, carrying out a targeted strike is deemed feasible only when the lives of innocent civilians are not endangered: "We only authorize a strike if we have a high degree of confidence that innocent civilians will not be injured or killed" (Brennan 2012). While minimising the damages inflicted on the population is a desirable and noble aim, discerning whether a strike poses a threat to civilian lives necessarily limits the counterinsurgent's ability to neutralise enemy combatants. As reported by a U.S. drone operator during an interview, in his unit "as many as 85 per cent to 90 per cent of requested strikes were rejected . . . because of concerns over civilian casualties or a lack of information about the target" (Smith 2016).

These concerns do not characterise the approach to targeted killings followed by counterinsurgents that opt for the brutalisation paradigm. As soon as personalised intelligence on subversive individuals becomes available, the counterinsurgent can effectively retaliate against anyone sharing ties with the rebellion—regardless of whether an individual has taken part in the anti-incumbent activity. Defined by scholars of strategic studies as "deterrence by punishment," this approach to civilian victimisation leverages the incumbent's capability of accomplishing acts of personalised retribution to minimise the insurgents' potential for collective action (Snyder 1960). In coercion-centred COIN operations, civilian victimisation falling under the deterrence paradigm comes in two configurations. The first form of selective deterrence specifically targets civilians believed to be insurgents or active sympathisers. Entailing what Kalyvas defines as the "personalisation" of retribution, such strategy presupposes "an intention to ascertain individual guilt" made possible by the availability of information submitted by local agents and civilian collaborators (Kalyvas 2006: 142). In the North Caucasian republic of Ingushetia, the security forces routinely murder individuals allegedly involved in the insurgency to deter the population from taking part in the rebellion. As accounted by a local eyewitness, the security forces often perform the killing in front of large crowds to better spread the message:

> There were gunshots [at the market] and everyone ran from all directions to see [what was happening] . . . The boy was killed like a quail . . . People were crowding here, and they could see this whole picture.
>
> (Human Rights Watch 2008: 36)

In Syria, pro-regime forces in government-controlled areas routinely carry out summary executions of individuals believed to be part of insurgent networks. As reported by a relative of a man murdered by the security forces, the regime liquidates suspected rebels in broad daylight to warn the population that the same destiny awaits those who side with the insurgency:

> In the morning soldiers came and took my nephew. They did not say anything and only had a quick look around the house . . . When the soldiers left, after about 15 minutes, we looked from the window and saw him lying face down in the pool of blood in the street.
>
> (Amnesty International 2012: 15–16)

By selectively killing civilians suspected of taking part in subversive activities, the counterinsurgent heightens the individual costs of taking part in rebellion whilst increasing the benefits of compliance, as "civilians can be relatively certain that cooperation can be exchanged for the right to survive" (Weinstein 2007: 18). In signalling that unaligned individuals will not be targeted for as long as they remain neutral, the incumbent provides the population with a transparent normative system that incentivises compliance in return for survival (Downes 2007).

The practice of targeting selected individuals is complemented and reinforced by the systematic targeting of entire groups of people who share a personal connection with individual non-compliers according to a "guilt by association" logic (Kalyvas & Kocher 2007: 188). Addressed in the literature on civil war as a "retributive" form of collective punishment, this strategy builds upon the previous to maximise the potential costs of engaging in insurgent activity (Souleimanov & Siroky 2016). By holding friends and relatives accountable for the deeds of single insurgents, the counterinsurgent pressures enemy fighters into either capitulating or defecting in the hope to spare their loved ones from certain retribution. Endowed with credibility by the government's willingness to carry out collective punishments—including looting, rape, torture, forcible disappearance, and public execution—this strategy signals to potential insurgents and civilian supporters that disobedience comes at a very high price. The history of COIN warfare is rich in episodes of violent retribution executed in a "deterrence by punishment" configuration. In Ancient China, the ruling dynasties would punish subversive individuals by wiping out their entire families: "the guiltless fathers and mothers, wives and children, and brothers and sisters were co-adjudicated and exterminated" (Teng 1977: 142). Collective punishments constituted a standard operating procedure for the German forces deployed in central Italy during the Second World War. In the effort to punish the partisans and deter the population from aiding the insurgents hiding in the mountains, the *Schutzstaffel* (SS) divisions exterminated entire villages inhabited by friends and relatives of local insurgents. Entered in the collective memory of the Italian nation, large-scale massacres of civilians, such as the one that occurred in the area of Monte Sole in September 1944, epitomise the profound psychological impact that the threat of collective punishment engenders amongst potentially hostile populations (Olsen 1968). More recently, the Israeli security forces engaged in the systematic demolition of the houses of alleged Palestinian terrorists to degrade their organisational infrastructure, rendering thousands of families homeless and sending a clear message to the rest of the Palestinian people (Hatz 2020). By sidestepping the restrictions placed by hearts and minds proponents on the use of force in COIN operations, a counterinsurgent can fully capitalise on its military prowess to achieve key strategic objectives.

Second pillar: political legitimacy

Considered by U.S. military doctrines as "principle of war" and "main objective" of asymmetric conflicts, the quest for establishing legitimacy— defined by Hammond as "the population's acceptance of a set of rules or an authority" (Hammond 2010: 69)—has become a leading objective in COIN operations (U.S. Gov. 2014a: 1–19; U.S. Gov. 2018: A-4). Rooted in the intellectual legacy of Max Weber, who famously described the state as the only entity capable of monopolising "the legitimate use of physical force" over the national territory, the assumption that strengthening government legitimacy inhibits civil conflict lays at the basis of any population-centric COIN campaign (Weber 1946: 77; Kitzen 2017; Nachbar 2012). In securing the population's obedience, conceptualised in Weberian terms as ranging from passive acquiescence to active adherence, the incumbent discourages individuals from engaging in violent activities at the expense of the rebels' potential for exacerbating social unrest. Although promoting legitimacy is of the utmost importance in COIN operations, the irreconcilable values of the hearts and minds approach and the brutalisation paradigm have engendered contrasting variants of state-building operations.

According to the *FM 3–07: Stability Operations*, spreading liberalism and consolidating democracy are "the hallmarks of a well-functioning government" (U.S. Gov. 2016: II-6). Hearts and minds advocates consider social contracts as unsatisfactory unless the majority of the population actively takes part in the political process. The importance of fostering political participation as a way to defeat insurgency resides at the core of some of the most influential studies exploring the dynamics of COIN warfare. While Kilcullen (2010: 5) considers "political legitimacy and respect for the rule of law" as the "operational imperatives" of any COIN operation, Kitzen (2012) and Bell (2011) profess that inclusive governments are more likely to succeed in restoring peace and addressing the rebellion's root causes. According to Chiarelli and Michaelis, this approach is the sole capable of providing long-term stability to the war-torn country of Iraq: "the purple index finger, proudly displayed, became a symbol of defiance and hope. The Iraqi people proved to the world their willingness to try democracy in whatever unique form it evolves" (Chiarelli & Michaelis 2005: 4).

Yet, a recently growing literature underlines that the "free and fair election" formula scores consistently low success rates in insurgency-affected societies. As explained by U.S. Army Major Wiechnik, the Western approach to nation-building is based on the assumption, rooted in a poor understanding of political legitimacy, that "if one creates democratic systems, including legislatures and executives filled by elected representatives, the population will embrace democracy" (Wiechnik 2012: 23). Yet, evidence from recent attempts of nation-building provides a dissatisfying picture of bottom-up political processes. The first round of elections held in Iraq in January 2005 underscores the perils of premature democratic processes. With the sectarian tensions between the disenfranchised Sunni communities and the empowered Shia minority rising by the day, the elections served as a catalyst for the spike in ethno-sectarian violence experienced across the country during the following year. While all major Sunni political parties called for a national boycott to protest against the elections' perceived illegitimacy, the leading Shia cleric, Grand Ayatollah Ali al-Sistani, issued a fatwa urging the Shias to meet their "religious duty" and cast their vote (Tavernise 2005). The credibility of the U.S. as a guarantor of political legitimacy amongst the Sunni communities was further damaged by the relative ease with which the insurgents deterred the population from showing up at polling stations in Sunni governorates. As reported by Walker and Clark, the elections "took place under conditions of extreme insecurity and political turmoil that made it all but impossible for every eligible voter freely to make a choice" (Walker &

Clark 2005: 187). Deprived of any meaningful representation in the new, heavily sectarian Shia government, the disenfranchised Sunni communities became exposed to Shia-led sectarian attacks—with many yielding to the propaganda of Sunni insurgent groups. In Iraq, the poorly engineered electoral processes sponsored by the external COIN force engendered an inflammatory, rather than a de-escalatory effect on intra-communal and insurgent violence. As argued by Etzioni, developing resilient democratic institutions can be a "slow, arduous process that is nearly impossible for outsiders to direct" (Etzioni 2012: 66). The upsetting experience of the 2005 elections held in Iraq resembles the disappointing results obtained in Afghanistan during the previous year. The first presidential election in the history of Afghanistan was held in the midst of an ongoing conflict between a Coalition of foreign forces and an enraging grassroots insurgency. While the high turnout at the polls and the relatively peaceful election of Hamid Karzai as the first president of Afghanistan were initially deemed a success of democracy over lawlessness, a closer scrutiny reveals important drawbacks brought about by the bottom-up electoral processes. The Coalition's inability to provide a minimum degree of protection at the polling stations confirmed the Afghan warlords as the true, undisputed rulers outside the territories directly controlled by the Western forces. As stated by the then-director of Human Rights Watch Brad Adams, the electoral process lacked credibility amongst the communities living in contested or insurgent-controlled areas: "Many voters in rural areas say the militias have already told them how to vote, and that they're afraid of disobeying them" (Human Rights Watch 2004). These assertions found widespread confirmation in recent studies carried out on the viability of democracy in Afghanistan. While Smith argues that the 2004 elections spawned a malfunctioning democratic tradition in which "election have essentially become a means of securing and freezing in place a fragile political settlement" (Smith 2012: 4), Greene (2017) went as far as defining the process of promoting democracy in traditionally non-democratic societies as "pathological counterinsurgency." Alongside Gawthorpe's (2017) assessment of local elections in civil war as hardly ever successful, empirical evidence demonstrates that building political legitimacy from the ground-up might not be as effective as conventionally surmised.

In contrast to the hearts and minds approach, political efforts performed according to the brutalisation paradigm are ingrained in the Hobbesian principle stating that, without an authoritative power regulating society, life is "solitary, poor, nasty, brutish, and short" (Hobbes 1965[1651]: 97). As a condition of civil war exposes individuals to "continual fear and danger of violent death" (Ibid.), frightened populations will legitimise any authority, even if coercively imposed, capable of ending the hostilities and restoring public order (Millen 2007). Despite legitimacy-building processes suffering from being "under-theorized, over-generalised, and misunderstood," scholars confirmed that repression plays a role of primary importance for the consolidation of sovereignty in territories contested with grassroots insurgent movements (Gawthorpe 2017: 849; Gerschewski 2018). By intensifying and then capitalising on the population's incessant demand for stability, the incumbent secures its control over a population terrorised into acquiescence by the promise of a draconian retribution against anyone found in league with the rebels (Levitsky & Way 2010). This oppressive approach to legitimacy-building characterised the COIN operations carried out by Fascist Italy during the pacification campaign in Libya (1922–1931). After having wiped out the indigenous resistance to the Italian invasion force, Rome enforced its direct control over the country through a combination of exemplary punishments and invasive policing activities (Rochat 2005: 12–14). Intimidated into submission by the Fascist surveillance apparatus, the population in Tripolitania started collaborating with the Italian authorities. As underscored by Dotolo in his study of Italy's COIN operation in North Africa, this strategy facilitated the enforcement of the local governorate's legitimacy, as people knew that "the war was not against [the] population per se; it was solely against the rebels" (Dotolo 2015: 176).

Although repression never ceases constituting the bedrock of the legitimacy-building efforts carried out under the brutalisation paradigm, counterinsurgents often sponsor general elections as a means to provide an additional layer of legitimacy to the incumbent. As explained by Kendall-Taylor and Frantz (2014), nowadays' incumbents "mimic" democratic practices, such as periodically held elections and multi-party systems, in an effort to manipulate popular dissent and appease the population's desire for transitioning towards a more inclusive political system. Previous research conducted on the role played by state-controlled political elections shows that counterinsurgents are keen to capitalise on electoral processes to consolidate the status quo (Croissant & Hellman 2018), signal regime invincibility (Seeberg 2014), minimise the recurrence of anti-government protests (Günay & Dzihic 2016), and redistribute material incentives to co-opted elites (Bray et al. 2019). While the proliferation of pseudo-democratic institutions should not conceal the fact that state-sponsored elections are "hardly more than a façade" utilised to magnify the regime's control over society, empirical research confirms that electoral processes can assist incumbents in garnering consensus and reduce the levels of pro-insurgent popular support (Soest & Grauvogel 2017: 292). In El Salvador, the 1984 presidential elections backed by the U.S. Government strengthened the local government at the expense of the insurgents, who partially lost their credibility in spite of the clearly fraudulent electoral process. As explained by Bello and Herman, the elections marked a watershed for the pro-incumbent camp:

> even though electoral conditions were as unfavourable to a free election as in 1982, the government program of symbolism and propaganda prevailed as before, with turnout and rebel disruption heavily featured in the media and the election hailed as a triumph of democracy.
>
> (Bello & Herman 1984: 856)

This effect was further amplified by the negative reputation obtained by the insurgents. Attempts aimed at sabotaging the elections were seen by the population as an attack against everyone's political rights and thus impacted negatively on the insurgents' reputation:

> [the guerrillas] took away ID cards from voters, an implicit threat. They refused to declare a ceasefire on election day, and some military actions were close enough to polling places so as to give the impression of an attack on the elections.
>
> (Spence & Vickers 1994: 6)

The 2014 Syrian presidential elections held amidst an ongoing civil war constitute a recent instance of electoral processes utilised by an incumbent to consolidate its authority and delegitimise the insurgents. Winning by a landslide against a largely non-existent political opposition, President Bashar al-Assad reconfirmed its ascendancy over Syria, "convincingly" signalling to the population that the rebels stood no chance of overthrowing a resilient authority enjoying widespread popular support (Anderson 2015; Sly & Ramadan 2014).

When compared with the legitimacy-building endeavours performed by counterinsurgents following the hearts and minds approach, it is clear that the brutalisation paradigm promises faster results at lower political costs. If hearts and minds textbooks advance complex procedures meant to garner legitimacy via popular will, the brutalisation paradigm prioritises the employment of long perfected instruments of coercive dominion to secure the regime's uncontested supremacy over the local population. In contrast with the theoretical precepts of the hearts and minds approach, evidence from the experience of counterinsurgents following the brutalisation paradigm confirms that spreading democracy is but one of many ways to engender political stability—and not amongst the most effective ones either.

Third pillar: economic development

In 2006, U.S. General David Petraeus published a list of 14 observations refined from the U.S. experience against the Iraqi insurgency. Prominent among these remarks is the maxim asserting that "money can be more important than ammunition" (Petraeus 2006). Reiterated in the *Commander's Guide to Money as a Weapons System*, which states that "money is one of the primary weapons used by war-fighters to achieve successful mission results," this notion has found widespread validation in academia and considerable application in contemporary theatres of COIN warfare (U.S. Gov. 2009b: 1; Cohen et al. 2006; Bodnar & Gwinn 2010; Berman et al. 2011). Defined by Donley as "the provision of sufficient basic services, infrastructure, and economic essentials to garner popular support and engender government legitimacy" (Donley 2016: 103), economic development constitutes a twofold instrument that counterinsurgents dispose to counter the enemy's propaganda and promote societal resilience (U.S. Gov. 2009a: 17; Choharis & Gavrilis 2010). Although these strategic objectives are typical of all COIN campaigns, mutually incompatible visions for a functional post-conflict environment induce counterinsurgents to operationalise contrasting templates for economic development operations.

Most of the aid programmes sponsored by the hearts and minds approach are ingrained in opportunity-cost theories suggesting that individuals can be discouraged from supporting insurgents if provided with unconditional access to unwavering livelihood opportunities (SIGAR 2018a; German Gov. 2010: 1). Embracing the premise that civil unrest endures where poverty rates are high (Fearon 2008), Western COIN doctrines recommend boosting economic prosperity to lower the rebels' potential for recruitment and increase the levels of pro-incumbent popular support (U.S. Gov. 2014a: 10–10; Weintraub 2016). To reduce the insurgency's potential for growth, practitioners following the hearts and minds approach inject local markets with large-scale monetary spending meant to re-activate the economy and incentivise the rise of a capitalistic, autonomous society. Although financial expenditures are channelled into a multitude of different projects aimed at maximising the chances of economic recovery (Zürcher 2017), the *FM 3–24* considers the promotion of small private enterprises and community-based projects as the key objective of development operations (U.S. Gov. 2006: 5–17). As concluded by Kilcullen, Mills, and Oppenheimer, "ensuring the smooth operation of the [local] market" triggers a "reinforcing cycle of recovery and prosperity" that contributes to solving the conflict's economic root causes (Kilcullen et al. 2011: 106).

The past 20 years of uninterrupted COIN operations in Iraq and Afghanistan required an impressive commitment of "monetary ammunition" in the form of economic development schemes. Since the fiscal year 2002, the U.S. Government has provided approximately $35.06 billion to support social and development aid programmes implemented in Afghanistan (SIGAR 2020: 130). This trend is likely to continue in the foreseeable future. On top of the $400 million destined to Afghanistan for the sole fiscal year of 2020, the DoD requested an additional $115 million in economic support and development funds for Iraq (U.S. Gov. 2019b: 12; 46). Given the monumental scale of these programmes, it would be reasonable to expect a noticeable return on the overall COIN effort. Yet, evidence on the effectiveness of economic development schemes in hearts and minds operations is scant and largely disappointing. Drawing upon a large dataset of geolocated violent incidents in Afghanistan, Sexton finds that civilian aid causes "a significant *increase* in insurgent violence" when allocated to areas in which the insurgents dispose of the violent means necessary to countervail the incumbent's presence (Sexton 2016: 731). These findings are consistent with previous studies carried out on the effects brought upon by economic aid on insurgent activity. After carrying out a large-scale randomised field experiment across 500 Afghani villages, Beath and colleagues have found that economic aid has a violence-dampening effect *only* in areas where the presence of security forces prevents the insurgents from orchestrating attacks: "development programs are more effective in preventing the spread of violence, rather than in reducing the level of violence" (Beath et al. 2012: 21). Similarly, Berman and colleagues find that, in Iraq, aid spending reduced the levels of insurgent violence only after a large increase in the number of U.S. troops occurred with the so-called "Surge" in January 2007 (Berman et al. 2011). Additionally, studies carried out on the effectiveness of economic development programmes show that small, flexible projects with a limited budget are correlated with more statistically significant decreases in insurgent activity (Berman et al. 2013; Adams 2015). Nevertheless, research carried out on the effectiveness of economic aid in conflict zones raised surprisingly little interest from Western practitioners. The fact that in Afghanistan "the areas to receive the largest amount of money are also the least secure and most violent" speaks volumes about the Western states' perception of what works and what does not in a given COIN operation (Johnson et al. 2012: 84).

While evidence from recent quantitative studies circumscribes the effectiveness of development programmes to government-controlled areas, an in-depth inquiry into the ways in which aid spending is administered reveals additional sets of deficiencies. The endemic presence of corruption, coupled with a chronic deficit in security, has simultaneously been a chief cause of ineffectiveness and an important source of income for insurgent groups. In Afghanistan, experts estimate that "approximately 10 percent of the Pentagon's logistics contracts—hundreds of millions of dollars—consists of payments to insurgents" (Roston 2009). Considering that the lack of security in Afghanistan's roads exposes the convoys to the risk of insurgent attacks, local trucking companies utilise the money offered by the counterinsurgent to buy a safe passage from the insurgents: "Our firm knowingly pays [rebel] thieves to ensure the safety of our cargo" (Schifrin 2010). In Iraq, local insurgents have seized millions of dollars by infiltrating the oil refineries reopened thanks to the money provided by the U.S. Government in the form of economic development funds. In an interview released in 2006, the then-finance minister of Iraq Ali Allawi confirmed that the insurgents were reaping between 40 and 50 per cent of all oil-smuggling profits in the country: "the insurgents are involved at all levels" (Worth & Glanz 2006). This was confirmed 2 years later by a U.S. Captain in command of several platoons stationed at the Baiji refinery, one of the most important industrial sites in Iraq: "It's the money pit of the insurgency" (Oppel 2008).

The accidental financing of insurgent groups is further aggravated by the widespread distrust of economic development schemes felt amongst the populations that receive the aid. This is important because, as explained by Böhnke and Zürcher, the success of economic development programmes must also be assessed in relation to the effects engendered within the target group: "the impact of development aid is largely in the eyes of the beholder" (Böhnke & Zürcher 2013: 427). When development aids are implemented without giving much thought to the local population's needs and customs, the counterinsurgent's efforts to win hearts and minds might not only result in failure but also backfire. As explained by an Afghan tribesman, engaging in economic development programmes without the explicit approval of the tribal leaders can be perceived as a great insult to the community's dignity and reputation:

> money can't win hearts and minds. If you give an Afghan a great meal but insult him, he will never come again. But if you treat him with respect but only give him a piece of bread, he will be your friend forever.
>
> (Wilder & Gordon 2009)

In other cases, the counterinsurgent might be spending substantial amounts of resources without realising that the population's perceptions and convictions can hardly be changed. In a series of interviews conducted with the residents of small Kenyan villages threatened by Islamist insurgents, the interviewees manifested diffidence—or outright frustration—towards the projects implemented by the U.S. Forces: "All goes well as long [as] the civil affairs presence is what the community wants, but they can turn against them. There is nothing long-lasting to build a relationship on" (Bradbury & Kleinman 2010: 59).

Promoting local entrepreneurial activities and community-based projects can contribute to generate communal wealth and reduce the numbers of unemployed people forced to join the insurgency out of economic necessity. Yet, evidence from recent theatres of COIN warfare shows that "even when well-implemented, foreign aid can be extraordinarily disruptive" or incidentally counterproductive (Choharis & Gavrilis 2010: 39). Additionally, the emergence of a self-sufficient economy has the side effect of decreasing the government's ability to exercise control over society. In Afghanistan, the prospect of privatising and/or liquidating state-owned enterprises was received with widespread scepticisms from the local population. According to a report by the Special Inspector General for Afghanistan Reconstruction (SIGAR), local residents were concerned that moving away from a state-led economy would bring even more instability at the local level: "most Afghans were unenthusiastic about privatization because they still looked to the state to be the lead economic actor" (SIGAR 2018b: 32). While reducing the population's dependency on state-owned enterprises can contribute to containing the spread of the insurgency by promoting societal resilience, an overly enthusiastic abuse of economic development aid as a tool of COIN warfare can often bring about counterproductive results.

These problems do not affect the brutalisation paradigm, which espouses a diametrically opposed vision for economic development operations. Instead of identifying the population as the recipient of economic benefits, the brutalisation paradigm recommends conveying material resources into programmes aimed at buying off selected social elites, such as defected rebel chiefs, religious leaders, and influential businessmen (Hazelton 2017: 91). Defined by Gerschewski (2013: 22) as "the capacity to tie strategically relevant actors to the regime," co-optation is a relatively inexpensive stabilisation strategy designed to provide local power holders with the instruments necessary to enforce order and generate compliance. Despite co-optation being criticised as prone to backfire when the government fails to satisfy the elite's expectations (Brenner 2015), a burgeoning literature sustains that co-opting the only individuals within society possessing "the talent, resolve, and social status to organise economic, political, or military activities that will antagonise violent insurgents" is the optimal way to administer state-controlled economies during periods of social turmoil (Moyar 2011: 6; Wilson & Akhtar 2019; Raleigh & Dowd 2018).

In COIN operations, the practice of hindering the emergence of a private sector responds to the imperative of reducing the resources that insurgents extract from their civilian supporters. Falling under the category of strategies defined by Leites and Wolf (1970: 36) as "input-denial," this approach rests on coercive measures "to diminish the supply of human and material resources available for rebel use" (Mason 1996: 75). By depriving suspected non-compliers of the possibility to find employment in state-owned enterprises, the incumbent forces the population into choosing between two mutually exclusive options: people either stop collaborating with the insurgents in return for economic gains or continue disobeying the authority at the risk of incurring into severe punishment (Moore 1995). If reducing the population's possibilities to free-ride—that is to say, enjoying the collective benefits provided by the incumbent whilst continuing supporting the opponent—is not an uncommon practice amongst counterinsurgents following the hearts and minds approach (Dugan & Chenoweth 2012; Evans 2014), what is unique to the brutalisation paradigm are the extreme punishments inflicted upon disobeyers. By depriving insurgent supporters of the means necessary to provide for their families, the incumbent raises to an unacceptable level the costs of defying the authority—a threat endowed with credibility by the incumbent's access to personalised information on actual and potential disobeyers. Prospective rebels are put in front of a choice: providing allegiance to the incumbent in return for a life of relative comfort or supporting the rebellion at the risk of seeing their relatives sentenced to a life of hunger and affliction. As free-riding is precluded by the ubiquitous presence of government informants within society, individuals are strongly encouraged to accept the government's offer and turn their back against the insurgency. Deprived of fresh recruits and crucial resources, isolated insurgents become easy targets for the government's security forces.

In doing so, the incumbent achieves three main objectives. First, forcing the population to economically depend on the state weakens the political opposition and undermines the civil society's potential for collective action (Holdo 2019: 447). Second, co-opting selected leadership figures deepens existing social cleavages and/or exacerbates intra-communal tensions that prevent the insurgents from accessing important pools of resources and manpower. In Yemen, the government has for years followed this "divide and conquer" strategy to contain the spread of the Houthi insurgency (2004–ongoing). As explained by a local sheik during an interview, pitting one tribe against the other precluded the insurgents from creating a united popular front against the government:

> For more than forty years the state has meddled in the [governorate of] al-Jawf in order to weaken the tribes by sowing discord among them . . . It endeavoured to exacerbate issues of feud and revenge among them in order to widen the gap between the tribes and prevent any convergence between them.
>
> (Brandt 2017: 258)

Third, forcing entire communities to rely on the state for their security and economic well-being compels the population into perceiving the insurgents as a top-priority threat. During the Algerian War of 1954– 1962, the exploitation of local feuds allowed the French to recruit up to 180,000 local tribesmen as indigenous forces collectively known as *Harkis*. As explained by Gortzak, the *Harkis* proved to be a loyal and determined fighting force against an insurgency considered a threat to the locals' security and economic well-being: "*Harkis* [were made up of individuals] who were driven by a personal thirst for revenge against the [Front of Libération Nationale] FLN . . . or who derived significant material, social, or political benefits from cooperation with the French" (Gortzak 2009: 323).

A revealing illustration of this shortcut pattern to economic development is offered by Murtazashvili's analysis of the Uzbek regime's survival strategy. To retain power over society, the Uzbek government assigned to co-opted powerholders the task of operating the state-owned welfare system. By forcing people to depend on the state for healthcare, education, food subsidies, and employment, the government compelled the population into passive acquiescence and raised to an extreme level the potential costs of supporting the rebels (Murtazashvili 2012: 86). Building on Murtazashvili's work, Ucko (2016) further revealed that the Chinese government adopted a similar approach to defeat the insurgency in the province of Xinjiang. To placate the rebellion, Beijing co-opted several opposition leaders and hired a large part of the unoccupied population to carry out public works meant to improve the local inhabitants' life conditions. Paid by the government to build public infrastructures designed to satisfy the local population's most pressing needs, many individuals chose to accept the government's offer over living a life of severe hardship, condemning the weakened insurgency to slowly wither away under the pressure exercised by the Chinese security forces (Odgaard & Nielsen 2014: 540).

Instead of devolving considerable resources in earning the entire population's gratitude, counterinsurgents following the brutalisation paradigm utilise development aid to win over a carefully selected minority capable of optimising state-sponsored recovery programmes. By forcing the population to rely on the state for satisfying essential economic needs, a counterinsurgent leaves the population with no other option than forsaking the insurgents and submitting to the incumbent. This aggressive, but nevertheless effective approach to economic development allows the regime to consolidate its dominion over society without having to compromise with the population in return for its acquiescence.

Overarching objective: success in population-centric COIN operations

After following mutually incompatible pathways throughout the execution of the COIN campaign (see Table 3.1), the orthodox and shortcut patterns converge in the last operational step, which entails the achievement of long-lasting success. Yet, success in population-centric COIN operations constitutes an elusive concept, difficult to define and even harder to operationalise. This is because, as Bartholomees suggests in his essay titled "Theory of Victory," success in a war largely corresponds to "an assessment, not a fact or condition" (Bartholomees 2008: 26). This notion holds particular relevance for COIN operations, a typology of armed conflict in which 80% of the effort is attributable to socio-political activities and the remaining 20% of military engagement can rarely deliver victory on its own (Zellen 2012: 132; Galula 2006[1964]: 63). The difficulty of conceptualising success in COIN warfare is epitomised by the *FM 3–24*, which does not provide a definition of victory and openly specifies that "following the principles and imperatives [provided in the manual] does not guarantee success" (U.S. Gov. 2006: 1–20).

As official military doctrines neglect to determine what success in COIN warfare looks like, the analysis must turn once again to Galula, who dedicated a small section of his book to identifying the parameters for victory in population-centric COIN operations. According to Galula, victory comprises two principal sets of objectives. The first one, derived from the Clausewitzian tradition embraced by enemy-centric proponents, includes the destruction of the insurgency's fighting force, its loss of morale, and its admission of the defeat by giving up its intentions (Clausewitz 1984: 234). But because insurgency cannot be defeated solely by military means, neglecting to pursue political solutions would allow the insurgents to reconsolidate their organisational infrastructure and wage new seasons of protracted warfare. Building on the enemy-centric definition of success, Galula further specified that long-lasting victory is preconditioned on the

Table 3.1 Principal divergences between orthodox and shortcut patterns

	Orthodox pattern	*Shortcut pattern*
Intelligence penetration	Persuasive	Coercive
Information operations	Qualitative approach	Quantitative approach
Use of force	Minimum	Unrestrained
Political legitimacy	Bottom-up	Top-down
Economic development	Private market economy	State-controlled economy

"permanent isolation of the insurgent from the population, isolation not enforced upon the population but maintained *by and with* the population" (Galula 2006[1964]: 54, emphasis added).

Although Galula's theorisations brought the political sphere to the forefront of population-centric COIN operations, his excessively vague definition of success led scholars and practitioners to misinterpret the concept of victory and bring it to an extreme end. Experts increasingly consider success as unachieved until the incumbent eliminates all the "broad public grievances" on which the insurgency fed to gain momentum (Shemella 2015: 64; Connable & Libiki 2010: 154). While addressing the root causes of rebellion decreases the potential for an insurgent's comeback, achieving this ideal-type objective constitutes an extremely arduous task that only a handful of counterinsurgents at most would be able to accomplish. With victory conceptualised in these terms, counterinsurgents have little to no chances of seeing their efforts being repaid by full-fledged success. Originated from a misconception of the tenets of population-centric COIN warfare, this idealistic framing of victory threatens to obscure Galula's description of success as the (re)building of "a political machine" capable of exercising uncontested control over society (Galula 2006[1964]: 95). Put otherwise, Galula underscores that establishing control over the local population—and not addressing the full range of popular grievances—constitutes the linchpin of population-centric COIN victory. This interpretation is confirmed by Kilcullen, who remarks that COIN warfare can be assimilated to a competition between the state and the rebels in which whoever does better at "establishing a predictable, consistent, wide-spectrum normative system of control is most likely to dominate" the population, overpower its opponent, and prevail in the armed struggle (Kilcullen 2013: 126).

Accepting the twofold interpretation of success as provided by Galula and shared by Kilcullen, this book extrapolates from recent quantitative research conducted on the sources of success in asymmetric warfare the parameters necessary to empirically assess the effectiveness of population-centric COIN endeavours. The first contribution to this book's definition of success is derived from Zhukov's research on the determiners of victory in COIN warfare. Measured by the frequency of insurgent attacks and degree of popular support received by the rebels, the "disruption of an insurgency's ability to sustain its operations" is considered by Zhukov as the narrowest threshold for COIN victory, assimilable to the enemy-centric propositions utilised by Galula as part of his interpretation of success (Zhukov 2008: 7). The second set of criteria is extracted from *Paths to Victory*, a RAND study focused on assessing the roots of victory in COIN operations by drawing upon one of the most empirically rich datasets on asymmetric conflicts currently available. Embodying the second part of Galula's definition of success, *Paths to Victory* considers the incumbent as "unambiguously" triumphant if the government stayed in power throughout the armed struggle, the country remained intact, and no major concessions were granted to the insurgents (Paul et al. 2013: 17). Taken together, these two studies supply the parameters necessary to operationalise Galula's interpretation of victory and apply it to the outcomes obtained with both the orthodox and shortcut patterns (see Table 3.2). In identifying a standardised list of criteria suited to assess the performances of both the hearts-and-minds and brutalisation ventures, this book provides a common normative standard of evaluation, so far lacking in the COIN literature, necessary to estimate the degree of success achieved by *any* counterinsurgent waging a population-centric COIN warfare.

The choice of utilising a single set of criteria to assess the effectiveness of hearts and minds and brutalisation efforts might result as puzzling. The brutalisation paradigm represents a "poor COIN concept" which "sits uneasily" within the view of population-centric COIN warfare presented in Western military doctrines (Paul et al. 2013: 108–109). By affirming that the same criteria for success can be applied to hearts and minds *and* brutalisation endeavours, the present book questions the intellectual foundations underpinning the hearts and minds approach, further suggesting that checklists of "bad COIN practices" might, in fact, constitute perfectly valid playbooks for victory in population-centric COIN operations (Paul & Clarke 2016: 3). Should a brutalisation-centred COIN campaign fulfil the criteria for success in population-centric COIN warfare, the theoretical foundations underpinning the hearts and minds approach would be problematised, and the common wisdom on the (in)effectiveness of several COIN practices should undergo a significant re-evaluation.

Table 3.2 Common normative standard of evaluation for population-centric COIN engagements

First set of criteria: The "enemy-centric" legacy	Second set of criteria: The "population-centric" cornerstone
Substantial decline in the frequency of insurgent attacks	Government stayed in power throughout the armed struggle
Substantial decline in levels of pro-insurgent popular support	The country remained intact (no secession occurred)
	No major concessions granted to the insurgents

References

Abouzeid, Rania. *No Turning Back: Life, Loss and Hope in Wartime Syria* (New York: W. W. Norton, 2018). ISBN: 978-0-393-35678-6.

Adams, Greg. "Honing the Proper Edge: CERP and the Two-Sided Potential of Military-Led Development in Afghanistan." *The Economics of Peace and Security Journal* 10(2), 2015: 53–60. https://doi.org/10.1080/14742837.2019.1597698.

Amnesty International. *Deadly Reprisals: Deliberate Killings and Other Abuses by Syria's Armed Forces*, 2012. Available at: www.amnesty.org/en/documents/MDE24/041/2012/en/.

Anderson, Tim. "America's 'Dirty War on Syria': Bashar al Assad and Political Reform." *Global Research*, 4th December 2015. Available at: www.globalresearch. ca/americas-dirty-war-on-syria-bashar-al-assad-and-political-reform/5492661.

Armistead, Leigh. *Information Operations: Warfare and The Hard Reality of Soft Power* (Washington, DC: Potomac Books, 2004).

Asal, Victor; Deloughery, Kathleen & Sin, Steve. "Democracy, Inclusion, and Failure in Counterinsurgency." *Foreign Policy Analysis* 13, 2017: 913–930. https://doi.org/10.1093/fpa/orw018.

Ashour, Omar. "Sinai's Insurgency: Implications of Enhanced Guerrilla Warfare." *Studies in Conflict & Terrorism* 42(6), 2019: 541–558. https://doi.org/10.1080/1057610X.2017.1394653.

Bartholomees, Boone J. "Theory of Victory." *Parameters*, Summer 2008: 25–36.

Beath, Andrew; Christia, Fotini & Enikolopov, Ruben. "Winning Hearts and Minds Through Development? Evidence from a Field Experiment in Afghanistan." *The World Bank*, Policy Research Working Paper 6129, 2012.

Bell, Colleen. "Civilianising Warfare: Ways of War and Peace in Modern Counterinsurgency." *Journal of International Relations and Development* 14, 2011: 309–332. https://doi.org/10.1057/jird.2010.16.

Bello, Walden & Herman, Edward S. "U.S.-Sponsored Elections in El Salvador and the Philippines." *World Policy Journal* 1(4), 1984: 851–869.

Berman, Eli; Felter, Joseph H.; Shapiro, Jacob N. & Troland, Erin. "Modest, Secure, and Informed: Successful Development in Conflict Zones." *American Economic Review: Papers & Proceeding* 103(3), 2013: 512–517. https://doi.org/10.1257/aer.103.3.512.

Berman, Eli; Shapiro, Jacob N. & Felter, Joseph H. "Can Hearts and Minds Be Bought? The Economics of Counterinsurgency in Iraq." *Journal of Political Economy* 119(4), 2011: 766–819.

Bodnar, Seth & Gwinn, Jeremy. "'Monetary Ammunition' in a Counterinsurgency." *Parameters* 40(3), 2010: 91–102.

Böhnke, Jan R. & Zürcher, Christoph. "Aid, Minds and Hearts: The Impact of Aid in Conflict Zones." *Conflict Management and Peace Science* 30(5), 2013: 411–432. https://doi.org/10.1177/0738894213499486.

Bradbury, Mark & Kleinman, Michael. *Winning Hearts and Minds? Examining the Relationship Between Aid and Security in Kenya* (Medford: Feinstein International Center, 2010).

Brandt, Marieke. *Tribes and Politics in Yemen: A History of the Houthi Conflict* (Oxford: Oxford University Press, 2017). ISBN: 9780190911775.

Bray, Laura A.; Shriver, Thomas E. & Adams, Alison E. "Framing Authoritarian Legitimacy: Elite Cohesion in the Aftermath of Popular Rebellion." *Social Movement Studies* 18(6), 2019: 682–701. https://doi.org/10.1080/14742837.2019.159 7698.

Brennan, John O. *The Ethics and Efficacy of the President's Counterterrorism Strategy*, 30th April 2012. Available at: www.wilsoncenter.org/event/the-efficacy-and-ethics-us-counterterrorism-strategy.

Brenner, David. "Ashes of Co-Optation: From Armed Group Fragmentation to the Rebuilding of Popular Insurgency in Myanmar." *Conflict, Security & Development* 15(4), 2015: 337–358. https://doi.org/10.1080/14678802.2015.1071974.

Byman, Daniel. "'Death Solves All Problems': The Authoritarian Model of Counterinsurgency." *Journal of Strategic Studies* 39(1), 2016: 62–93. https://doi.org/10.1080/01402390.2015.1068166.

Caesar, Julius. *The Gallic War* (London: William Heinemann, 1919).

Chiarelli, Peter W. & Michaelis, Patrick R. "Winning the Peace: The Requirement for Full-Spectrum Operations." *Military Review*, July–August 2005: 4–17.

Choharis, Peter C. & Gavrilis, James A. "Counterinsurgency 3.0." *Parameters*, Spring 2010: 34–46.

Clausewitz, Von Carl. *On War* (London: Everyman's Library, 1984). ISBN: 0-691-05657-9.

Cohen, A. A. *Galula: The Life and Writings of the French Officer Who Defined the Art of Counterinsurgency* (Santa Barbara: Praeger Security International, 2012).

Cohen, Eliot; Crane, Conrad; Horvath, Jan & Nagl, John. "Principles, Imperatives, and Paradoxes of Counterinsurgency." *Military Review*, March–April 2006: 49–53.

Collings, Deirdre & Rohozinski, Rafal. S*hifting Fire: Information Effects in Counterinsurgency and Stability Operations* (Carlisle: U.S. Army War College, 1st Edition, 2005).

Connable, Ben & Libiki, Martin C. *How Insurgencies End* (Santa Monica: RAND Corporation, 1st Edition, 2010). ISBN: 978-0-8330-4952-0.

Copeland, Daryl & Potter, Evan H. "Public Diplomacy in Conflict Zones: Military Information Operations Meet Political Counter-Insurgency." *The Hague Journal of Diplomacy* 3, 2008: 277–297.

Croissant, Aurel & Hellman, Olli. "Introduction: State Capacity and Elections in the Study of Authoritarian Regimes." *International Political Science Review* 39(1), 2018: 3–16. https://doi.org/10.177/01925121170066.

Donley, Patrick H. "Economic Development in Counterinsurgency: Building a Stable Second Pillar." *Joint Force Quarterly* 81(2nd Quarter), 2016: 102–111.

Dotolo, Frederick H. "A Long Small War: Italian Counterrevolutionary Warfare in Libya, 1911 to 1932." *Small Wars & Insurgencies* 26(1), 2015: 158–180. https://doi.org/10.1080/09592318.2014.959765.

Douhet, Giulio. *The Command of Air* (Washington, DC: Office of Air Force History, 1983[1942]). ISBN: 0-912799-10-2.

Downes, Alexander B. "Desperate Times, Desperate Measures: The Causes of Civilian Victimisation in War." *International Security* 30(4), 2006: 152–195. Available at: www.jstor.org/stable/4137532.

Downes, Alexander B. "Draining the Sea by Filling the Graves: Investigating the Effectiveness of Indiscriminate Violence as a Counterinsurgency Strategy." *Civil Wars* 9(4), 2007: 420–444. https://doi.org/10.1080/13698240701699631.

Dugan, Laura & Chenoweth, Erica. "Moving Beyond Deterrence: The Effectiveness of Raising the Expected Utility of Abstaining from Terrorism in Israel." *American Sociological Review* 77(4), 2012: 597–624. https://doi.org/10.1177/0003122412450573.

Etzioni, Amitai. "The Folly of Nation Building." *The National Interest*, July–August 2012: 60–68.

Evans, Ryan. "'The Population Is the Enemy': Control, Behaviour, and Counter-Insurgency in Central Helmand Province, Afghanistan." In *The New Counter-Insurgency Era in Critical Perspective*, edited by Jones, David M.; Gventer, Celeste W. & Smith, M. L. R (Hampshire: Palgrave Macmillan, 2014). https://doi.org/10.1057/9781137336941.

Fearon, James D. "Economic Development, Insurgency, and Civil War." In *Institutions and Economic Performance*, edited by Elhanan, Helpman (Cambridge, MA: Harvard University Press, 1st Edition, 2008). ISBN: 978-0-674-03077.

Galula, David. *Counterinsurgency Warfare: Theory and Practice* (London: Praeger Security International, 10th Edition, 2006[1964]), 30–41. ISBN: 0-275-99269-1.

Gant, Jim. *One Tribe at a Time: A Strategy for Success in Afghanistan* (Los Angeles: Nine Sisters Imports, 2009).

Gawthorpe, Andrew J. "All Counterinsurgency is Local: Counterinsurgency and Rebel Legitimacy." *Small Wars & Insurgencies* 28(4–5), 2017: 839–852. https://doi.org/10.1080/09592318.2017.1322330.

German Government. *Preliminary Basics for the Role of Land Forces in Counterinsurgency* (Cologne: German Army Office, NATO Unclassified, 2010).

Gerschewski, Johannes. "Legitimacy in Autocracies: Oxymoron or Essential Feature?" *Perspectives on Politics* 16(8), 2018: 652–665. https://doi.org/10.1017/S1537592717002183.

Gerschewski, Johannes. "The Three Pillars of Stability: Legitimation, Repression, and Co-Optation in Autocratic Regimes." *Democratization* 20(1), 2013: 13–38. https://doi.org/10.1080/13510347.2013.738860.

Getmansky, Anna. "You Can't Win If You Don't Fight: The Role of Regime Type in Counterinsurgency Outbreaks and Outcomes." *The Journal of Conflict Resolution* 57(4), 2013: 709–734. Available at: www.jstor.org/stable/24545614.

González, Roberto J. "Beyond the Human Terrain System: A Brief Critical History (and A Look Ahead)." *Contemporary Social Science* 15(2), 2018: 1–14. https://doi.org/10.1080/21582041.2018.1457171.

Gortzak, Yoav. "Using Indigenous Forces in Counterinsurgency Operations: The French in Algeria, 1954–1962." *Journal of Strategic Studies* 32(2), 2009: 307–333. https://doi.org/10.1080/01402390902743415.

Greene, Samuel R. "Pathological Counterinsurgency: The Failure of Imposing Legitimacy in El Salvador, Afghanistan, and Iraq." *Third World Quarterly* 38(3), 2017: 563–579. https://doi.org/10.1080/01436597.2016.1205439.

Grossman, Dave. *On Killing: The Psychological Cost of Learning to Kill in War and Society* (New York: Back Bay Books, 1996). ISBN: 0-316-33000-0.

Günay, Cengiz & Dzihic, Vedran. "Decoding the Authoritarian Code: Exercising 'Legitimate' Power Politics through the Ruling Parties in Turkey, Macedonia and Serbia." *Southeast European and Black Sea Studies* 16(4), 2016: 529–549. https:// doi.org/10.1080/14683857.2016.1242872.

Hammond, James W. "Legitimacy and Military Operations." *Military Review*, 2010: 68–79.

Hatz, Sophia. "Selective or Collective? Palestinian Perceptions of Targeting in House Demolition." *Conflict Management and Peace Science* 37(5), 2020: 515–535. https://doi.org/10.1177/0738894218795134.

Hazelton, Jacqueline L. "The 'Hearts and Minds' Fallacy: Violence, Coercion, and Success in Counterinsurgency Warfare." *International Security* 42(1), 2017: 80–113. https://doi.org/10.1162/ISEC_a_00283.

Herpen, Van Marcel. *Putin's Propaganda Machine: Soft Power and Russian Foreign Policy* (Lanham: Rowman & Littlefield, 2016). ISBN: 978-1-4422-5362-9.

Hobbes, Thomas. *Leviathan* (London: Oxford University Press, 1965[1651]).

Holdo, Markus. "Cooptation and Non-Cooptation: Elite Strategies in Response to Social Protest." *Social Movement Studies* 18(4), 2019: 444–462. https://doi.org/10.1080/14742837.2019.1577133.

Hultquist, Philip. "Is Collective Repression an Effective Counterinsurgency Technique? Unpacking the Cyclical Relationship between Repression and Civil Conflict." *Conflict Management and Peace Science* 34(5), 2017: 507–525. https://doi.org/10.1177/0738894215604972.

Human Rights Watch. *Afghanistan: Warlords Threaten Integrity of Election*, 29th September 2004. Available at: www.hrw.org/news/2004/09/29/afghanistan-warlords-threaten-integrity-election.

Human Rights Watch. *"As If They Fell from the Sky": Counterinsurgency, Rights Violations, and Rampant Impunity in Ingushetia*, June 2008. Available at: www. hrw.org/report/2008/06/24/if-they-fell-sky/counterinsurgency-rights-violations-and-rampant-impunity.

Jackson, Brian A. "Counterinsurgency Intelligence in a 'Long War': The British Experience in Northern Ireland." RAND Corporation, *Military Review*, January–February 2007: 74–85.

Johnson, Gregory; Ramachandran, Vijaya & Walz, Julie. "CERP in Afghanistan: Refining Military Capabilities in Development Activities." *PRISM* 3(2), 2012: 81–98.

Johnston, Patrick B. & Urlacher, Brian R. "Explaining the Duration of Counterinsurgency Campaigns." 5th April 2021. Available at: http://patrickjohnston.info/materials/duration.pdf.

Jonas, Susanne. "Guatemala: Acts of Genocide and Scorched-Earth Counterinsurgency War." In *Centuries of Genocide: Essays and Eyewitnesses Accounts*, edited by Totten, Samuel; Parsons, William S. & Charny, Israel W. (New York: Routledge, 4th Edition, 2013). ISBN: 0-415-94430-9.

Jones, Seth G. *Waging Insurgent Warfare: Lessons from the Viet Cong to the Islamic State* (London: Oxford University Press, 2017). ISBN: 9780190600860.

Joseph, Paul. "'Soft Power' Does Not Always Mean 'Smart Power': An Investigation of Human Terrain Teams in Iraq and Afghanistan." *Palgrave Communications*, December 2016.

Kalyvas, Stathis N. *The Logic of Violence in Civil War* (Cambridge: Cambridge University Press, 1st Edition, 2006). ISBN: 978-0-511-22508-6.

Kalyvas, Stathis N. "The Paradox of Terrorism in Civil War." *The Journal of Ethics* 8, 2004: 97–138.

Kalyvas, Stathis N. "Wanton or Senseless? The Logic of Massacres in Algeria." *Rationality and Society* 11(3), 1999: 243–285.

Kalyvas, Stathis N. & Kocher, Matthew A. "How Free is Free Riding in Civil Wars? Violence, Insurgency, and the Collective Action Problem." *World Politics* 59(2), January 2007: 177–216. https://doi.org/10.1353/wp.2007.0023.

Kendall-Taylor, Andrea & Frantz, Erica. "Mimicking Democracy to Prolong Autocracies." *The Washington Quarterly* 37(4), 2014: 71–84. https://doi.org/10.1080/0163660X.2014.1002155.

Kilcullen, David. *The Accidental Guerrilla: Fighting Small Wars in the Midst of a Big One* (New York: Oxford University Press, 1st Edition, 2009). ISBN: 978-0-19-536834-5.

Kilcullen, David. *Counterinsurgency* (London: C. Hurst & Company, 2010). ISBN: 978-0-19-973748-2; 978-0-19-973749-9.

Kilcullen, David. *Out of the Mountains: The Coming Age of The Urban Guerrilla* (London: Hurst, 1st Edition, 2013). ISBN: 978-0-19-973750-5.

Kilcullen, David. "Three Pillars of Counterinsurgency." *U.S. Government Counterinsurgency Conference, Washington, DC*, 28th September 2006.

Kilcullen, David; Mills, Greg & Oppenheimer, Jonathan. "Quiet Professionals." *The RUSI Journal* 156(4), 2011: 100–107.

Kitzen, Martijn. "Close Encounters of the Tribal Kind: The Implementation of Cooption as a Tool for De-escalation of Conflict—The Case of the Netherlands in Afghanistan's Uruzgan Province." *Journal of Strategic Studies* 35(5), 2012: 713–734. https://doi.org/10.1080/01402390.2012.706972.

Kitzen, Martijn. "'Legitimacy is the Main Objective': Legitimation in Population-Centric Counterinsurgency." *Small Wars & Insurgencies* 28(4–5), 2017: 853–866. https://doi.org/10.1080/09592318.2017.1322331.

Kusher, Barak. *The Thought War: Japanese Imperial Propaganda* (Honolulu: University of Hawai'i Press, 2006). ISBN: 978-0-8248-2920-9.

LeGree, Larry. "Thoughts on The Battle for the Minds: IO and COIN in the Pashtun Belt." *Military Review*, September–October 2010: 21–32.

Leites, Nathan & Wolf, Charles. *Rebellion and Authority: An Analytic Essay on Insurgent Conflicts* (Santa Monica: Rand Corporation, 1970). ISBN: 8410-0909-0.

Levitsky, Steven & Way, Lucan A. *Competitive Authoritarianism: Hybrid Regimes After the Cold War* (Cambridge: Cambridge University Press, 2010). ISBN: 978-0-511-90226-0.

Lyall, Jason. "Do Democracies Make Inferior Counterinsurgents? Reassessing Democracy's Impact on War Outcomes and Duration." *International Organization* 64(1), 2010: 167–192. https://doi.org/10.1017/S00208 18309990208.

Lyall, Jason. "Does Indiscriminate Violence Incite Insurgent Attacks? Evidence from Chechnya." *Journal of Conflict Resolution* 53(3), 2009: 331–362. https://doi.org/10.1177/0022002708330881.

Lyall, Jason & Wilson III, Isaiah. "Rage against the Machines: Explaining Outcomes in Counterinsurgency Wars." *International Organization* 63(1), 2009: 67–106. Available at: www.jstor.org/stable/40071884.

Mansoor, Peter R. *Baghdad at Sunrise: A Brigade Commander's War in Iraq* (London: Yale University Press, 2008). ISBN: 978-0-300-14069-9.

Marcucci, Megan. *Breaking the Silence: The Story of the Ixil Maya of Union Victoria During the Guatemalan Civil War* (Fairfield: Sacred Heart University, History Undergraduate Publications, 2017).

Marszalek, John F. *Sherman: A Soldier's Passion for Order* (Carbondale: Southern Illinois University Press, 2007 [1993]).

Mason, David T. "Insurgency, Counterinsurgency, and the Rational Peasant." *Public Choice* 86(1/2), 1996: 63–83. Available at: www.jstor.org/stable/30027069.

McKelvey, Tara. *Monstering: Inside America's Policy of Secret Interrogation and Torture in the Terror War* (New York: Basic Books, 2007). ISBN: 978-0-78671-776-7.

Millen, Raymond. "The Hobbesian Notion of Self-Preservation Concerning Human Behavior during an Insurgency." *Parameters*, Winter 2006/2007: 4–13.

Miroiu, Andrei. "Wiping Out 'The Bandits': Romanian Counterinsurgency Strategies in the Early Communist Period." *The Journal of Slavic Military Studies* 23(4), 2010: 666–691. https://doi.org/10.1080/13518046.2010.526021.

Moore, Will H. "Rational Rebels: Overcoming the Free-Rider Problem." *Political Research Quarterly* 48(2), 1995: 417–454. Available at: www.jstor.org/stable/449077.

Moyar, Mark. "Development in Afghanistan's Counterinsurgency: A New Guide." *Orbis Operations*, March 2011. Available at: https://smallwarsjournal.com/documents/development-in-afghanistan-coin-moyar.pdf.

Munoz, Arturo. *U.S. Military Information Operations in Afghanistan: Effectiveness of Psychological Operations 2001–2010* (Santa Monica: RAND Corporation, 2012).

Murtazashvili, Jennifer. "Coloured by Revolution: The Political Economy of Autocratic Stability in Uzbekistan." *Democratization* 19(1), 2012: 78–97. https://doi.org/10.1080/13510347.2012.641295.

Nachbar, Thomas B. "Counterinsurgency, Legitimacy, and the Rule of Law." *Parameters*, Spring 2012: 27–38.

Nagl, John A. *Learning to Eat Soup with a Knife: Counterinsurgency Lessons from Malaya and Vietnam* (Chicago: The University of Chicago Press, 13th Edition, 2005). ISBN: 0-275-97695-5.

NATO. *COMISAF's Counterinsurgency Guide* (Kabul, Afghanistan: COMINSAF/CDR USFOR-A, 2010b). Available at: http://glevumassociates.com/doc/speech_COMISAF-COIN-Guidance.pdf.

NATO. *ISAF Commander's Counterinsurgency Guidance* (Kabul, Afghanistan: HQ ISAF, 2010a). Available at: www.nato.int/isaf/docu/official_texts/counterinsurgency_guidance.pdf.

Nye, Joseph S. "Information Warfare versus Soft Power." *The Strategist*, 12th May 2017. Available at: www.aspistrategist.org.au/information-warfare-versus-soft-power/.

Odgaard, Liselotte & Nielsen, Galasz Thomas. "China's Counterinsurgency Strategy in Tibet and Xinyang." *Journal of Contemporary China* 23(87), 2014: 535–555. https://doi.org/10.1080/10670564.2013.843934.

Olsen, Jack. *Silence on Monte Sole* (New York: Putnam, 1st Edition, 1968). ISBN: 0213177943.

Oppel, Richard A. "Iraq's Insurgency Runs on Stolen Oil Profits." *The New York Times*, 16th March 2008. Available at: www.nytimes.com/2008/03/16/world/middleeast/16insurgent.html.

Paul, Christopher & Clarke, Colin P. *Counterinsurgency Scorecard Update* (Santa Monica: RAND Corporation, 1st Edition, 2016). ISBN: 978-0-8330-9262-5.

Paul, Christopher; Clarke, Colin P.; Grill, Beth & Dunigan, Molly. *Paths to Victory: Lessons from Modern Insurgencies* (Santa Monica: RAND Corporation, 1st Edition. 2013). ISBN: 978-0-8330-8054-7.

Pechenkina, Anna O.; Bausch, Andrew W. & Skinner, Andrew K. "How Do Civilians Attribute Blame for State Indiscriminate Violence?" *Journal of Peace Research* 56(4), 2019: 545–558. https://doi.org/10.1177/0022343319829798.

Peters, Ralph. *Wars of Blood and Faith: The Conflicts That Will Shape the Twenty-First Century* (Mechanicsburg: Stackpole Books, 2007). ISBN: 978-0-8117-0274-4.

Petraeus, David H. "Learning Counterinsurgency: Observations from Soldiering in Iraq." *Military Review*, January–February 2006: 2–12.

Raleigh, Clionadh & Dowd, Caitriona. "Political Environments, Elite Co-Option, and Conflict." *Annals of the American Association of Geographers* 108(6), 2018: 1668–1684. https://doi.org/10.1080/24694452.2018.1459459.

Roca, Raimundo R. "Information Operations during Counterinsurgency Operations: Essential Option for a Limited Response." *Athena Intelligence Journal* 3(1), 2008. ISSN: 1998–5237.

Rochat, Giorgio. *Le Guerre Italiane 1935–1943: Dall'Impero d'Etiopia alla Disfatta* (Torino: Giulio Einaudi Editore, 1st Edition, 2005). ISBN: 88-06-16118-0.

Roston, Aram. "How the US Funds the Taliban." *The Nation*, 11th November 2009. Available at: www.thenation.com/article/archive/how-us-funds-taliban/.

Schifrin, Nick. "Report: U.S. Bribes to Protect Convoys Are Funding Taliban Insurgents." *ABC News*, 22nd June 2010. Available at: https://abcnews.go.com/WN/Afghanistan/united-states-military-funding-taliban-afghanistan/story?id=10980527.

Seeberg, Merete B. "State Capacity and the Paradox of Authoritarian Elections." *Democratization* 21(7), 2014: 1265–1285. https://doi.org/10.1080/13510347.2014.960210.

Sexton, Renard. "Aid as a Tool against Insurgency: Evidence from Contested and Controlled Territory in Afghanistan." *American Political Science Review* 110(4), 2016: 731–749. https://doi.org/10.1017/S0003055416000356.

Shemella, Paul. "Strategy and Outcome: What Does Winning Look Like?" In *The Future of Counterinsurgency: Contemporary Debates in Internal Security Strategy*, edited by Lawrence, Cline E. & Shemella, Paul (Santa Barbara: Praeger Security International, 2015). ISBN: 978-1-4408-3300-7.

SIGAR (Special Inspector General for Afghanistan Reconstruction). *Oversight of U.S. Spending in Afghanistan* (Washington, DC, 2018a).

SIGAR. *Private Sector Development and Economic Growth: Lessons from the U.S. Experience in Afghanistan* (Washington, DC, 2018b). Available at: www.sigar. mil/pdf/lessonslearned/SIGAR-18-38-LL.pdf.

SIGAR. *Quarterly Report to the United States Congress* (Washington, DC, April 2020). Available at: https://reliefweb.int/sites/reliefweb.int/files/resources/2020-04-30qr.pdf.

Sly, Liz & Ramadan, Ahmed. "Syrian Election Sends Powerful Signal of Assad's Control." *The Washington Post*, 3rd June 2014. Available at: www.washing tonpost.com/world/middle_east/syrian-election-sends-powerful-signal-of-assads-control/2014/06/03/16876fca-eb2a-11e3-b98c-72cef4a00499_ story.html.

Smith, Josh. "U.S. Drone Pilots Defend Tactics as Afghans Question Civilian Toll." *Reuters*, 22nd December 2016. Available at: www.reuters.com/article/us-afghan-istan-drones-idUSKBN14B0X0.

Smith, Rupert. "Interview with General Sir Rupert Smith." *International Review of the Red Cross* 88(864), 2006: 719–727.

Smith, Rupert. *The Utility of Force: The Art of War in the Modern World* (New York: Alfred A. Knopf, 1st Edition, 2007).

Smith, Scott S. "The 2004 Presidential Elections in Afghanistan." In *Snapshots of an Intervention: The Unlearned Lessons of Afghanistan's Decade of Assistance (2001–11)*, edited by van Bijlert, Martine & Kuovo, Sari (Kabul: Afghanistan Analysts Network, 2012). Available at: www.afghanistan-analysts.org/wp-con-tent/uploads/downloads/2012/09/Snapshots_of_an_Intervention.pdf.

Snyder, Glenn H. "Deterrence and Power." *The Journal of Conflict Resolution* 4(2), 1960: 163–178.

Soest, von Christian & Grauvogel, Julia. "Identity, Procedures and Performance: How Authoritarian Regimes Legitimize their Rule." *Contemporary Politics* 23(3), 2017: 287–305. https://doi.org/10.1080/13569775.2017.1304319.

Souleimanov, Emil A. & Siroky, David S. "Random or Retributive? Indiscriminate Violence in the Chechen Wars." *World Politics* 68(4), 2016: 677–712. https://doi. org/10.1017/S0043887116000101.

Spence, Jack & Vickers, George. *Toward a Level Playing Field? A Report on the Post-War Salvadorian Electoral Processes* (Cambridge: Hemisphere Initiatives, 1994).

Statiev, Alexander. *The Soviet Counterinsurgency in the Western Borderlands* (New York: Cambridge University Press, 2010). ISBN: 978-0-521-76833-7.

Tavernise, Sabrina. "Top Shiite Cleric is Planning to Urge Iraqis to Back Char-ter." *The New York Times*, 23rd September 2005. Available at: www.nytimes. com/2005/09/23/world/top-shiite-cleric-is-planning-to-urge-iraqis-to-back-charter.html.

Teng, Y. S. "The Role of the Family in the Chinese Legal System." *Journal of Asian History* 11(2), 1977: 121–155. Available at: www.jstor.org/stable/41930248.

U.K. Government. *British Army Field Manual 1(10), Countering Insurgency* (Brit-ish Army, 2009).

U.S. Government. *Commander's Guide to Money as a Weapons System* (Washington, DC: Department of the Army, April 2009b).

U.S. Government. *COP-CO: FY 2020 Comprehensive Oversight Plan Overseas Contingency Operations.* Lead Inspector General (Washington, DC: Department of Defence, 2019b).

U.S. Government. *Counterinsurgency Guide* (Washington, DC: Interagency Counterinsurgency Initiative, 2009a).

U.S. Government. *Dictionary of Military and Associated Terms* (Washington, DC: Department of Defence, 2019a).

U.S. Government. *FM 3–24 MCWP 3–33.5 Counterinsurgency* (Washington, DC: Department of the Army, December 2006).

U.S. Government. *FM 3–24 MCWP 3–33.5: Insurgencies and Countering Insurgencies* (Washington, DC: Department of the Army, 2014a).

U.S. Government. *Joint Publication 3–0: Joint Operations* (Washington, DC: Department of the Army, 2018).

U.S. Government. *Joint Publication 3–07: Stability* (Washington, DC: Department of the Army, 2016).

U.S. Government. *Joint Publication 3–13: Information Operations* (Washington, DC: Department of the Army, 2014b).

Ucko, David H. "'The People are Revolting': An Anatomy of Authoritarian Counterinsurgency." *Journal of Strategic Studies* 39(1), 2016: 29–61. https://doi.org/10.1080/01402390.2015.1094390.

Valentino, Benjamin; Huth, Paul & Balch-Lindsay Dylan. "'Draining the Sea': Mass Killing and Guerrilla Warfare." *International Organisation* 58(2), 2004: 375–407. https://doi.org/10.1017/S0020818304582061.

Walker, Harold & Clark, Terence. "Elections in Iraq—30 January 2005: An Assessment." *Asian Affairs* 36(2), 2005: 181–191. https://doi.org/10.1080/03 068370500136247.

Weber, Max. "Politics as a Vocation." In *From Max Weber: Essays in Sociology*, edited by Gerth, H. H. & Mills, Wright (New York: Oxford University Press, 1946). ISBN: 0-415-17503-8.

Weinstein, Jeremy M. *Inside Rebellion: The Politics of Insurgent Violence* (Cambridge: Cambridge University Press, 2007). ISBN: 978-0-511-34864-8.

Weintraub, Michael. "Do All Good Things Go Together? Development Assistance and Insurgent Violence in Civil War." *The Journal of Politics* 78(4), 2016: 989–1002. https://doi.org/10.1086/686026.

Wiechnik, Stanley J. "Policy, COIN Doctrine, and Political Legitimacy." *Military Review*, November/December 2012: 22–30.

Wilder, Andrew & Gordon, Stuart. "Money Can't Buy America Love." *Foreign Affairs*, 1st December 2009. Available at: https://foreignpolicy.com/2009/12/01/money-cant-buy-america-love/.

Wilson, Chris & Akhtar, Shahzad. "Repression, Co-Optation and Insurgency: Pakistan's FATA, Southern Thailand and Papua, Indonesia." *Third World Quarterly* 40(4), 2019: 710–726. https://doi.org/10.1080/01436597.2018.1557012.

Worth, Robert F. & Glanz, James. "Oil Graft Fuels the Insurgency, Iraq and U.S. Say." *The New York Times*, 5th February 2006. Available at: www.nytimes.

com/2006/02/05/world/middleeast/oil-graft-fuels-the-insurgency-iraq-and-ussay.html.

Zellen, Barry S. *The Art of War in an Asymmetric World: Strategy for the Post-Cold War Era* (London: Continuum International Publishing Group, 2012). ISBN: 978-1-4411-5431-6.

Zenz, Adrian. "'Thoroughly Reforming Them Towards a Healthy Heart Attitude': China's Political Re-Education Campaign in Xinjiang." *Central Asian Survey* 38(1), 2019: 102–128. https://doi.org/10.1080/02634937.2018.1507997.

Zhukov, Yuri M. "Counterinsurgency in a Non-Democratic State: The Russian Example." In *The Routledge Companion to Insurgency and Counterinsurgency*, edited by Rich, Paul B. & Duyvesteyn, Isabelle (London: Routledge, 2011). ISBN: 978-0-415-56733-6.

Zhukov, Yuri M. "Evaluating Success in Counterinsurgency, 1804–2000: Does Regime Type Matter?" *Working Paper, 21st Convention of the International School on Disarmament and Research on Conflicts (ISODARCO)*, 2008. Available at: www.isodarco.it/courses/andalo08/paper/andalo08_Zhukov_paper.pdf.

Zhukov, Yuri M. "A Theory of Indiscriminate Violence." PhD Thesis, Harvard University, 2014.

Zürcher, Christoph. "What Do We (Not) Know About Development Aid and Violence? A Systematic Review." *World Development* 98, 2017: 506–522. https://doi.org/10.1016/j.worlddev.2017.05.013.

4 The brutalisation paradigm in Chechnya and elsewhere

Background of the conflict and narratives for Russia's counterinsurgency operations

"The Chechen wars of 1994–1996 and 1999–2009 were dramatic, vicious, and complex affairs," stated Galeotti (2014: 9) in his account of the vicissitudes experienced by the small, breakaway nation of Chechnya during its struggle for independence in the aftermath of the Soviet Union's collapse. Covering an area of approximately 17,300 square kilometres, roughly equivalent to the size of Montenegro, Chechnya is a land-locked autonomous republic located in the North Caucasus, a mountainous region included within the borders of the Russian Federation.

As with the secessionist wars that occurred in neighbouring post-Soviet republics, the Chechens' centuries-old ambition for national self-determination lays at the roots of the brutal conflicts that pitted a "small, but proud and warlike" population against a state striving to recover its territorial integrity after the disintegration of its multi-ethnic empire (Kipp 2001: 47). Triggered by the Chechen leadership's resolve in resisting the Kremlin's demands for capitulation, the First Chechen War represents an initial Russian attempt directed at crushing the Chechen uprising and enforcing Moscow's sovereignty over the rebellious republic. Anticipated by the Russian military leadership as a "bloodless blitzkrieg," the fighting quickly took an unexpected turn for the Russian forces, which entered Chechnya unprepared to face a resourceful enemy determined to wear out the invaders in a partisan warfare of ambushes and ruthless terrorism (Lapidus 1998: 20). Forced to withdraw after the insurgents re-obtained control of Grozny, Chechnya's capital, in August 1996, the Russians undertook an accelerated programme of military reforms intended to prepare the Army to confront and—this time—defeat the Chechen separatists (Baev 2003). The opportunity to take back the autonomous enclave came in 1999 when the incursion of a group of armed Chechen rebels in the neighbouring Republic of Dagestan gave

DOI: 10.4324/9781003188049-4

Moscow the pretext to launch a large-scale offensive against the *de facto* independent republic (Pain 2001: 11). After accomplishing an initial conventional phase aimed at conquering Grozny through an intensive campaign of artillery shelling and carpet bombings, the over 90,000-strong invasion force spread to the countryside and engaged the retreating rebels in a series of aggressive COIN operations (Oliker 2001: 122). Nine years after the start of the pacification efforts in Chechnya, the President of the Russian Federation Vladimir Putin announced to the State Council that, in the fight against the Chechen "terrorists," the Russian forces had prevailed, and that the insurgents had suffered a "decisive and crushing blow" from which they could never recover from (Kremlin 2008). While the scholarship generally confirms that the outcome of the Second Russian-Chechen War favoured the Russians, the nature of the COIN operations performed against the insurgency is highly contested in the literature.

One group of scholars interpreted the Russian COIN during the Second Russian-Chechen War as inherently enemy-centric. According to this scholarship, the Russian Army "singularly failed" to win the population's hearts and minds, as its irresistible "desire" for all-out conventional battle could not be reconciled with the protracted, low-intensity engagements characterising population-centric COIN operations (Hodgson 2003: 80; Janeczko 2012: 4). While these studies show that the Russians never ceased resorting to indiscriminate measures in the fight against the rebels, their conceptualisation of the Russian COIN as centred on physically exterminating the insurgents downgrades the strategic relevance placed by Moscow on seizing and maintaining long-term control over the local population. The incongruences between these approximative theoretical assumptions and the evidence emerging from empirical examinations of the Russian COIN characterise even widely acclaimed publications, as typified by Schaefer's book discussing the strategic measures implemented by Moscow to defeat the Chechen insurgency. While the author introduces the Russian COIN as a "clear example of enemy-centric approach in action" (Schaefer 2010: 7), his descriptions of the tactics utilised by the Russians to engender popular support for the local government mismatch the fundamentals of COIN warfare as outlined in enemy-centric narratives.

Although the scholarship consistently refrains from considering the Russian COIN as an offspring of the population-centric paradigm, a substantial body of research has acknowledged that enemy-centric narratives do not exhaustively explain the strategic rationale behind Russia's employment of a highly varied "instrumentarium" of military, political, economic, psychological, and civic measures. By interpreting the Russian COIN as "neither military-centric nor population-centric," this school of thought advances a "hybridity narrative" based on the belief that Russia "changed the variables in the Western standard counterinsurgency matrix to came up with her own autochthonous formula" (Grebennikov 2015: 75; Miakinkov 2011: 648). While the physical neutralisation of the insurgents continues to be considered as the main objective of the Russian military operations, the centre of analysis results shifted from the Chechen to the Russian domain of the COIN effort. In scrutinising the Russian media's coverage of the Second Russian-Chechen War, these studies demonstrated that Moscow, instead of securing popular support for the authority in Chechnya as recommended by the hearts and minds approach, utilised its non-kinetic tools to persuade the domestic public opinion that the Chechen "terrorists" constituted an intolerable top-priority security threat (Meakins 2017). Although these studies confirmed that the Second Russian-Chechen War can "certainly" be read through the lens provided by Western COIN textbooks, their conceptualisation of the Russian COIN as an enemy-centric endeavour contaminated by "reversed" hearts and minds precepts denotes a rather marginal consideration given to the political and economic efforts implemented *within* the Chechen territory (Kim & Blank 2013: 929).

While the proponents of enemy-centric and hybridity narratives enriched the available knowledge on the Second Russian-Chechen War, their accounts of the Russian COIN rely almost exclusively on empirical evidence and suffer from the mainstream literature's assumption of the brutalisation paradigm as incompatible with population-centric COIN templates. Lacking a solid theoretical understanding of the conceptual foundations of the population-centric paradigm, these studies ended up utilising inaccurate theoretical models to delineate the Russian COIN, with the consequence that inquiries conducted on the Second Russian-Chechen War present the same problematics affecting the wider literature on the brutalisation paradigm. In an effort to redirect the study of this conflict towards a more refined framework of analysis, this chapter advances the first empirical account of the Second Russian-Chechen War in which the Russian COIN is explicitly considered as an instance of population-centric COIN warfare in action.

Support base: intelligence penetration and information operations

In Chechnya, the weaponisation of fear never ceased being the "official state policy" of Moscow's intelligence services (Gilligan 2010: 70). While the Russians systematically terrorised the Chechen population for intelligence-gathering purposes, their operational toolkit underwent a notable evolutionary process throughout the execution of the COIN campaign.

The initial use of wantonly brutal intelligence techniques, including beatings, humiliations, rape, and torture, was dictated by the imperative of providing combat troops with constant streams of actionable intelligence (OCHA 2001). As combat units were suffering heavy casualties in the effort to crush the resistance (Kramer 2005: 214), the intelligence services were put under pressure to collect, collate, and distribute information necessary to pinpoint and liquidate high-value targets within the insurgency's ranks. To achieve this aim, the Russians rapidly set up a network of "filtration points"—clearinghouses in which suspected fighters and civilian supporters were detained and interrogated. Described as "cramped, filthy, and sordid," these detention centres were designed to force even the most determined insurgent sympathiser to reveal, under torture, the identities of other individuals associated with the rebellion (Human Rights Watch 2000: 38). Terrorised beyond imagination by the prospects of being "welcomed to [the] hell" of filtration camps, many Chechens ended up collaborating with the Russians to escape torture and death (Ibid.: 40). Although deplorable, these techniques served the purpose they were intended for. As one interviewee put it, "of course [the population] would collaborate . . . everywhere you could find a traitor . . . in every family there was one."[1]

While frightening the population into passive collaboration allowed the Russians to acquire much-needed intelligence, the system reliant on filtration camps was far from efficient. Not long after the start of the conflict, analysts such as Garwood (2002) and Peterson (2003) signalled that the insurgents were replenishing their ranks with new recruits determined to avenge their relatives tortured and murdered by Russian soldiers. Realising that combat troops were failing to make substantial progress, Moscow opted for a strategic turnaround and started replacing in 2004 the Russian servicemen with the so-called *Kadyrovtsy*, a local paramilitary force named after President Putin's chosen overlords for Chechnya—Akhmad Kadyrov and his son Ramzan. Largely composed of former insurgents persuaded to defect in exchange for a state pardon, the *Kadyrovtsy* could draw upon their intimate knowledge of the rebels' *modus operandi* and socio-cultural milieu to accurately identify and neutralise insurgent fighters and civilian supporters (Souleimanov 2015).

While incentivising rallied insurgents to share information on their former comrades-in-arms allowed the Russians to stem the tide of individuals joining the rebellion, forcing the defectors' families to rely on the government's successes for securing their own safety provided the COIN forces with massive inflows of high-quality intelligence. Fearing the insurgents' retaliation, the relatives of turncoat rebels started to "routinely" report any suspicious insurgent activity to the pro-government militias (Souleimanov & Aliyev 2015a: 697). Penetrating deep into society by leveraging on the population's fears, the Russians were able to consolidate "all-seeing, all-hearing" grids of local informants capable of tracking down the insurgents' movements.

As reported by Seierstad during her trip in Chechnya, "the strife . . . has entered a phase where streets have eyes, everyone watches everyone else, and anyone who doesn't denounce others is hiding something" (Seierstad 2008: 153). The psychological effects produced by this surveillance apparatus on the Chechen population hardly goes unnoticed by people travelling across Chechnya. While the relatives of one interviewee visiting his/her family were afraid to share their thoughts on the country's political situation as "there is always the risk of someone eavesdropping and reporting you to the authorities,"[2] the friends of another were telling him/her "to always watch [his/her] mouth, as walls have ears."[3]

If establishing a pervasive surveillance apparatus enabled the security forces to detect and react to early signs of insurgent activity, the tight control exercised over the sources of information available in Chechnya greatly facilitated the broadcasting of pro-government propaganda messages. Sealing off the Chechen territory from all non-authorised media sources constituted the first step of Moscow's information warfare campaign. By denying the rebels access to external sources of information and systematically destroying their systems of communication, including radio stations, cellular transmissions, and relay points, the Russians immediately seized the upper hand in the "war of words" waged against the insurgents' political machine (Thomas 2003: 211; Herd 2000). After accomplishing these tasks in the initial stages of the invasion, Moscow in 2001 launched a large-scale media offensive aimed at suffocating the enemy's propaganda infrastructure and saturating Chechnya with pro-government indoctrination material (Jaimoukha 2005: 232). Initially focused on "blaming the militants for everything" that the population suffered throughout the conflict, Russia's "quantitative approach" to IO was gradually refined as Moscow transformed the war from a struggle for independence into an inter-Chechen strife—a policy known as "Chechenisation" (Thomas 2005: 753; Ware 2009).

Under the Chechenisation agenda, the Russians exposed the population to campaigns of aggressive, brainwashing messages meant to stigmatise the rebels as radical Islamists committed to bringing Chechnya on the verge of a new civil conflict (Blank 2013). The effects obtained by this incessant disinformation on the socially polarised Chechen population should not be underestimated. According to several refugees, the state propaganda has been "a very effective tool for controlling people" in Chechnya.[4] As one interviewee put it:

> The regime is constantly utilising mass-media to influence the public's opinion. TVs, journals, lately social networks: everything contributes to brainwashing people into believing that the regime is doing good things for the Chechen population. [Ramzan Kadyrov] is always saying

that Putin saved Chechnya from terrorists, and that we should be grateful to the Russian Federation. It goes on every day . . . I think it really works. No freedom of choice, no freedom of speech. People are blindfolded, it is like living behind the iron curtain.[5]

Acting as a catalyst for societal polarisation, the regime propaganda dehumanised the rebels in the eyes of many Chechens, who "ended up believing in what they were told by the state"[6] and started perceiving the insurgents as a living cancer threatening the nation's security. Taken together, deep intelligence penetration and aggressive IO severely weakened the Chechen insurgency, disrupting the enemy's political apparatus whilst strengthening the government's ability to seize control over the local population.

The Russian experience with coercive intelligence and aggressive IO closely resembles the brutalisation-centred measures deployed by Soviet intelligence and IO officers in Afghanistan (1979–1989). Aware that a poor understanding of the Afghan socio-cultural terrain precluded their troops from connecting with the local population, the Soviets invested their resources in creating a widespread network of indigenous informants capable of identifying insurgent fighters and civilian supporters. Detachments of secret intelligence agents started recruiting "informants, guides, and other agents to expose the hiding places of the rebels in the towns and drew up plans of their houses," thus spreading suspicion and animosities amongst the population caught in the midst of the conflict (Mitrokhin 2002: 143). The Soviets spared no effort in their quest of penetrating deep into every Afghan village, even setting up an Afghan intelligence agency, known as *Khadamat-e Aetla'at-e Dawlati* (KHAD), to handle the recruitment and coordination of local informants. As reported by Hyman, KHAD scored substantial successes in gathering intelligence: "its lavish Soviet funding allows it to maintain an efficient network of some 20,000 paid informers in the provinces, as well as across the border in Pakistan" (Hyman 1984: 281). The effectiveness of the intelligence submitted by local collaborators was further strengthened by the systematic recruitment of local fighters and rallied insurgents in pro-Soviet indigenous forces. On the one hand, the co-optation of entire tribes situated in insurgent-controlled territories allowed the Soviets to set up rural militias with direct knowledge of the local socio-cultural terrain: "The [Soviet-controlled Democratic Republic of Afghanistan] DRA offered payments to tribes . . . along the Pakistani border or sought to exploit tribal antagonisms by recruiting a given tribe to curb the activities of a traditionally hostile neighbour" (Baumann 1993: 167). On the other hand, the Soviets manufactured or exacerbated animosities between groups of rebels to bring them on their side and thus acquire precious intelligence: "If the negotiations were successful the group officially surrendered its arms, only to take them up again at once, but this time in service of the government" (Dorronsoro 2005: 211). As reported by a Mujahideen leader, this strategy inflicted tremendous damages on the insurgency's infrastructure: "Why should the Soviets worry about killing Afghans if the Mujahideen do it for them?" (Girardet 2011[1985]: 129).

To drive an even larger wedge between the insurgents and the local population, the Soviets set up a massive IO infrastructure to simultaneously suffocate the Mujahideen's propaganda and spread the Communist ideology. As explained by Tchantouridzé, bombarding the population with pro-government indoctrination material allowed the Soviets to insulate the population from the insurgents' political messages:

> The Soviet propaganda, especially in the area of cinematic propaganda, proved to be tremendously popular among all Afghans . . . The Soviet-funded Agitprop teams, staffed by both Soviet servicemen and Afghan professionals, would visit an Afghan village, play popular music among the locals, distribute printed propaganda, display posters . . . Such events were well attended and welcomed by villagers.
>
> (Tchantouridzé 2013: 341)

While pro-Soviet Afghan media outlets consistently portrayed the Mujahideen as "fanatics disguised as Islam's defenders" who would "murder patriotic clergy and destroy mosques" (U.S. Gov 1987: 58), the Soviets attempted to "Sovietise" the Afghan society by teaching to thousands of local students the precepts of Communism:

> Ideological education forms an essential part of the curriculum at the university level. In the faculties of Kabul university, the courses of studies have been drastically changed and subjects like Revolutionary History of Workers, Dialectical Materialism, Scientific Sociology have been introduced.
>
> (Rizvi 1988: 84)

The tremendous effort that the Soviet put into intelligence penetration and IO further confirms that counterinsurgents following the brutalisation paradigm *do* place importance on non-kinetic activities. As explained by Robinson, this approach differs from hearts and minds precepts in its execution, not in its scope: "the difference between Soviets and Americans lies not so much in the volume of resources assigned to hearts and minds as to the manner in which they were integrated into overall operations" (Robinson 2010: 13–14).

First pillar: use of force

Subduing a population that considers bravery and self-sacrifice in war as its sacred values was no easy task for Moscow. Driven by what some interviewees defined as the "warrior soul" of the Chechen nation, at the start of the hostilities the population was firmly determined to fight the Russians until the last man. According to several interviewees, Chechens "never run away" from a fight,[7] and they are "always ready to protect their Motherland, no matter the cost."[8] This was confirmed by one eyewitness, who recalled that "we were willing to give our lives in fighting [the Russians]—we feared nothing. We were proud to sacrifice for our country's independence."[9] Confronted with such a fierce opponent, the Russians knew that victory would have been difficult to achieve without firstly breaking the "warrior soul" that made the Chechens coalesce in defence of their homeland.

To curtail the Chechens' potential for mass mobilisation, the Russians embarked on what Schaefer defined a "savage warfare" of indiscriminate, random violence (Schaefer 2010: 192). Premeditated and systematic, the intensive shelling of populated settlements signalled that the Russians were ready to exterminate entire villages in the effort to seize territorial supremacy. As declared by the then-Prime Minister Vladimir Putin, the Russians would have stopped in front of nothing to deprive the rebels of shelter and safe havens: "The bandits will be pursued wherever they are" (Gordon 1999). The profound societal shock produced by these brutalities is vividly recalled by those who directly witnessed this strategy in action. Initially, the all-out campaign of randomised shelling terrorised the population into paralysis, forcing people to seek shelter and stop providing logistical support to the rebels. As accounted by one interviewee, "I cannot express with words that kind of fear. In those moments, all you want to do is to run for cover."[10] The intensive shelling of populated areas was complemented with aggressive sweeps—infamously known as "*zachistka*." On the way to the border with Georgia, one refugee directly witnessed how the Russians executed *zachistkas*: "the soldiers were setting fire to many villages, burning houses with people still inside. They were also killing men on the spot and forcefully separating family members according to their gender."[11] Demoralised and traumatised, the vast majority of the population renounced taking part in the hostilities, with the consequence that the insurgents saw their support networks abruptly shrinking in the early stages of the war (Cohen 2014: 43). As one interviewee explained:

> during the First [Russian-Chechen] War, the Chechens fought as one entity . . . During the Second [Russian-Chechen War], however, the Russians destroyed our internal cohesion . . . This strategy broke our determination and prevented us from fighting as effectively as we did in the First War.[12]

Having terrorised the population into submission, the Russians gradually de-escalated the use of random attacks and, acting on the basis of more accurate intelligence submitted by captured and rallied insurgents, switched to more pinpointed forms of kinetic operations. Starting from early 2004, the selective targeting of rebels and civilian supporters by *Kadyrovtsy* units severely damaged the insurgency's operational infrastructure, which was gradually deprived of many high-ranking members liquidated during targeted operations performed by Russian special forces. The increased accuracy and frequency of these targeted raids exercised "tremendous pressure" on the insurgents to either capitulate or defect to the pro-Russian camp (Lyall 2010: 14). As underscored by Souleimanov, the selective targeting of insurgents and civilian supporters fragmented the insurgency "from within," leaving the rebels with no other option than accepting the government's amnesties and joining the fight against their former comrades-in-arms (Souleimanov 2017a: 35–39).

The systematic targeting of insurgents and civilian supporters has been complemented and reinforced by the extensive use of collective punishments against the insurgents' relatives in a "deterrence by punishment" configuration. Encapsulating a "guilt by association" logic, the practice of holding the insurgents' families responsible for attacks carried out against the security forces acted as a forceful deterrent for pro-insurgent collective action, heightening the costs of defying the authority well above any acceptable level of risk (Souleimanov & Siroky 2016). The severe consequences of upholding outlawed behaviours have been made unequivocally clear to the population, as epitomised by the following excerpt of a 2008 public discourse held on national television by the then-mayor of Grozny Khuchiev:

> If your relatives commit an act of evil, this evil will be brought upon you, your other family members and even your descendants . . . The evil perpetrated by your relatives from the woods will come back to your own houses and in the very near future everyone [of you] will feel it on your own back.
>
> (Human Rights Watch 2009: 24)

By openly warning the population that entire families would have been tortured and murdered for providing support to the rebels, the regime instilled in potential insurgent recruits the fear of disobeying the government's directives—a peril constantly reminded by exemplary punishments performed by the *Kadyrovtsy* against non-compliers. "It only takes few public punishments to scare the rest of the population into submission," noted an interviewee, as everyone in Chechnya is aware that "challenging the government constitutes a death sentence for you and your family."[13]

Despite the early successes of these brutalisation-centred measures, the systematic targeting of innocent civilians and the killing of insurgent relatives contributed to prolong the persistence of low-level insurgent violence across the Chechen territory. The practice of seeking blood revenge, defined by Chagnon (1988: 965) as "a retaliatory killing in which the initial victim's close kinsmen conduct a revenge raid on the members of the current community of the initial killer," is deeply rooted in the Chechen social texture and is considered by local Chechens as a man's highest social and moral obligation. As found by Souleimanov and Aliyev, the need to exact vengeance for the crimes committed by Russian servicemen and their indigenous allies "ultimately prompted thousands of Chechens who were initially apolitical . . . to resort to violence" (Souleimanov & Aliyev 2015b: 173). While the dynamics of blood-revenge have been a major motivational driver for the pro-insurgent violent mobilisation of local avengers, the regime's intensification of its kin killing policy effectively interrupted the cycle of killings motivated by the logic of blood-revenge—at least for the time being.

Fearful of the regime's open-ended threats against their relatives, many insurgents and potential avengers lowered their weapons, indefinitely postponing—or abandoning altogether—the commitment to take the fight against the government. Criticising the widespread climate of impunity enjoyed by the *Kadyrovtsy*, Politkovskaya concluded that the only feasible way for revenge-seekers to exact vengeance is to join a pro-government paramilitary unit:

> Everyone in Chechnya knows it. If you want to take blood revenge against someone, all you need to do is to get a job in Ramzan's unit: They will accept you, arm you, and give their blessing to your blood revenge.
>
> (The Jamestown Foundation 2003)

As remarked by numerous members of the Chechen diaspora, the population is acutely aware of the dilemmas faced by individuals desiring to retaliate against government representatives:

> you might want to take revenge against those who inflicted sufferings upon you and your family, but it is impossible to do so. Giving a pretext to the authorities is more than enough to trigger their reprisals, and the risk of losing your life and those of your loved ones in the process is too high.[14]

This was confirmed by another interviewee, who stated that "nobody would be so reckless as to put his family in grave danger to seek revenge against the government."[15] As long as the regime holds its iron grip over the population, blood revenge is unlikely to prompt individual avengers to seek membership into the insurgency's ranks. According to a former insurgent, only a period of chronic political instability could offer the rebels a chance to wage a new season of protracted warfare: "When President [Ramzan] Kadyrov will no longer be in office, many people might decide to exact blood revenge upon government officials and members of the *Kadyrovtsy* units."[16] Yet, Kadyrov's ability to maintain a nearly absolute control over the country's political institutions precludes individual avengers from "going to the woods" and seek violent retaliation.

Prevented from waging warfare in Chechnya, the remaining rebels were forced to either join insurgent groups in other areas of the Caucasus or confront the Russians in the Syrian battlegrounds of global Jihad (O'Loughlin & Witmer 2012; Ratelle 2016). By holding entire families accountable for the deeds of single insurgents, the Russians minimised the rebels' potential for mobilising the population in anti-incumbent collective actions. As emphasised by one interviewee, "those with a 'warrior soul' are no longer in Chechnya; they either died in the fighting . . . or joined the cause elsewhere—Syria, Iraq, Ukraine and other territories of the former Soviet space."[17] During the Chechen conflict, the extensive use of violence against civilians largely assisted the Russians in neutralising the insurgents and seizing control over the local population.

A similarly callous approach to the use of force in COIN operations was successfully implemented by the Italian Royal Army to suppress the Sanusi insurgency in Libya's Cyrenaica (1922–1931). In an effort to contain and defeat an insurgency that enjoyed the wholehearted support of its people, the Italians waged an incessant campaign of large-scale indiscriminate violence to frighten the population into passive submission and entice compliance at gunpoint. As explained by an Italian diplomat serving in Libya, the wantonly brutal campaign of repression was designed to break the Sanusi's spirit of resistance and clearly signal that there would have been no alternative but submission:

> Those fugitives running toward Egypt to take shelter are for us the best propaganda for the subdued populations. In particular, women tell about dread, misery, danger, anguishes incurred as well as the physical sufferings, sowing compassion and fright among the subdued, which, basically, were glad to have maintained their attitude of subdued.
>
> (Fasanotti 2020: 53)

Once established a military foothold in the region, the Italian military vice-governor for Cyrenaica Rodolfo Graziani engaged in the brutal and systematic liquidation of anyone found associated with the Sanusi resistance, setting up a "flying tribunal" to bring to trial and execute alleged insurgent combatants and civilian supporters (Baldinetti 2010: 47–48).

Yet, efforts aimed at selectively targeting individuals taking part in anti-incumbent activities were not enough to submit a population firmly determined to fight for its freedom. As accounted by Evans-Pritchard, the Italians could have never succeeded in persuading the population to stop providing material and logistical support to the rebels: "The *sottomessi* [subdued] and *ribelli* [rebels] alike were all Bedouin, jealous of each other and hostile to tribes other than their own, but united by blood and a common way of life—one faith, one speech, one law." (Evans-Pritchard 2008[1949]: 163). Faced with the population's resolute animosity, the Italian High Command changed strategy and ordered the collective punishment of entire families and tribes found guilty of providing support to the insurgents. While communities of *sottomessi* could consider themselves as safe from reprisals for as long as they remained subdued, no mercy would have been granted to the kinsmen of individuals in league with the insurgents. "I shall destroy all, men and things," announced General Pietro Badoglio, governor-general of Cyrenaica and Tripolitania (Gooch 2005: 1015). These were promises that the vice-governor Graziani was determined to turn into facts: "Let it be proclaimed that, following the desertion of five men from the Abadla el Bid tribe, I punished the entire collective of 80 tents with the confiscation of their cattle and the forcible transfer of the entire aggregate" (Graziani 1932: 105). Although the collective reprisals weakened the insurgency's cohesion, progress was slow, and a large fraction of the local population was resisting the Italians' campaign of systematic repression. Persuaded that nothing could be done to compel the reminder of the insurgency into capitulation, Graziani ordered the deportation of the entire population of Cyrenaica, between 85,000 and 100,000 tribesmen, to concentration camps. By exercising absolute control over the families of individuals serving within the insurgents' ranks, the Italians put the rebels in front of a grim choice: saving their loved ones by laying down their weapons or condemning their families to certain death. As recalled by a Sanusi tribesman, the relatives of insurgents were executed on a daily basis in the camps: "They were buried in mass graves. Fifty bodies a day, every day. We always counted them . . . People were either hanged or executed" (Del Boca 2005: 179). In Libya, brutalisation-centred measures—and not winning hearts and minds—ultimately proved to be the linchpin for success against one of the most determined insurgencies of the 20th century.

Second pillar: political legitimacy

In their monograph titled "*Russia's Restless Frontier*," Trentin and Malashenko defined Chechnya's political situation during the two wars' interlude as a slow "degeneration into anarchy" (Trentin & Malashenko 2004: 15).

Ever since the First Russian-Chechen War's aftermath, the Chechen resistance has been torn apart by a fierce competition for leadership between two principal camps—the nationalists, initially led by the elected president of the Chechen Republic of Ichkeria (ChRI) Aslan Maskhadov on one side, and the Islamists, guided by popular Jihadi warlords such as Shamil Basayev and Arbi Barayev on the other (Moore & Tumelty 2009). Refusing to accept Maskhadov's secular leadership as legitimate, the Islamists started challenging his authority, engaging in subversive activities to discredit the government as incapable of guaranteeing the enforcement of law and order. By the summer of 1997, the rivalry between the two factions was already reaching a tipping point, with armed confrontations occurring between Chechen troops and Jihadi fighters near Gudermes, Chechnya's second-largest city (Knysh 2007: 517). The 1999 Russian invasion further deepened the fissures that emerged within the insurgency's ranks. As explained by Toft and Zhukov, despite nationalists and Islamists coexisting in Chechnya "for more than 15 years, fighting the same enemy over the same terrain," their mutually incompatible objectives prevented the resistance from presenting a united front in the struggle against Russia (Toft & Zhukov 2015: 225).

It was on these premises that Moscow built its political scheme for Chechnya. The first and most important step in the consolidation process of a local pro-Russian government was the co-optation of Ahkmad Kadyrov—former *Mufti* of the CRhI and hard-lined separatist during the first conflict. Opposed to the spread of Salafi-Jihadism but also deeply aware of Russia's determination to crush the Chechen dream of independence, Kadyrov realised that defecting to the pro-Russian camp constituted his only chance to protect his clan from Moscow's destructive wrath (Russell 2011a). Designated as Moscow's endorsed candidate for the 2003 presidential elections, Kadyrov's rise from former insurgent to figure of authority of the new Chechen government signalled that Chechnya was not necessarily destined to remain what Politkovskaya (2003) described as a "small corner of hell." If a Western audience might consider Kadyrov a traitor, for many Chechens, his election constituted a long-awaited milestone in the process towards the normalisation of the country. As one interviewee remarked, "most people welcomed Kadyrov's decision, as it was clear to everyone that defeating the Russians was impossible, and that prolonging the hostilities would have only brought more sorrow upon our people."[18]

While Kadyrov's assassination by a group of rebels in 2004 threatened to derail Moscow's political scheme for Chechnya (Myers 2004), the appointment of Kadyrov's 27-year-old son Ramzan as his father's political successor prevented the pacification effort from reaching a dangerous dead-end. Defined by Russell (2008) as "the indigenous key to success in Putin's Chechenization strategy," Ramzan Kadyrov was able to rapidly consolidate his leadership position, leveraging on a mixture of charisma and intimidation to engender "genuine popular support" for his government (Dannreuther & March 2008: 98). To win the hearts of his suffering population, Kadyrov resorted to two main mechanisms. Outsmarting the Jihadists' political narrative by portraying himself as a restorer of the "morally declining" Chechen culture constituted an excellent strategy for gaining popularity among the older generations (Souleimanov 2006). By referring to Salafi-Jihadism as a devilish faith and vilifying its adherents as "enemies of Islam," Kadyrov drove an ideological wedge between radical insurgents and the vast majority of the population, which follows a more moderate form of Islam called "Sufism" (Smirnov 2006). The politicisation of religious and cultural values allowed Kadyrov to present himself as a guardian of God's law and preserver of traditions that Chechens care deeply about (Kurbanova 2011). This strategy permitted Kadyrov to secure the population's sympathy, as suggested by the comment provided by one interviewee, who stated: "what I like about Kadyrov is that he promotes and protects the customs of our people. He is a good leader, because he takes care of our cultural heritage."[19]

If preserving the nation's spiritual integrity contributed to stigmatise the rebels and increase popular support for the government, the rhetorical construction of Kadyrov's persona as the "saviour" of Chechnya granted him the sincere gratitude of many Chechens, who started admiring their young leader and building a personality cult around his figure. Treated with deferential respect by his admirers, "King Ramzan" leveraged on his popular appeal to indoctrinate the nation into considering his father and himself as bringers of peace and prosperity (The Independent 2007). Constantly hosted in popular TV shows and public events in Chechnya, Kadyrov has become the object of a nation-wide cult, as signified by the following passage extracted from a 2006 schoolchildren poetry contest titled "Ramzan—hero of our time":

> Praise to our radiant sun—Ramzan
> Mighty leader and fighting man
> To the faith and love and hope of Chechnya
> Grant a long life, O eternal Allah!
>   (quoted in Russell 2011b: 517).

The fame and admiration for Kadyrov extends well beyond Chechnya's borders. While a refugee considered Chechens as "better off with rather than without Kadyrov,"[20] another participant stated that "Chechens couldn't ask for a better president,"[21] stressing the fact that Kadyrov brought hope when before it was the only woe.

Particularly interesting is to note that the brutal crackdown on dissidents and the climate of repression enforced by the regime are not just accepted but also justified by many individuals who consider limiting the population's democratic freedoms as necessary to prevent people from threatening the current state of relative peace. As one interviewee put it:

> limiting the population's freedom of saying whatever it wants is a necessary evil, as people that have no restraints often do more harm than good with their words. This is especially true for Chechens: it is necessary for some to stay quiet and stop inciting others to think of violence as a viable solution for their problems."[22]

This opinion was shared by another refugee, who specified that

> people must understand that authorities deserve respect. . . . This is what Kadyrov does: he teaches people to stop shaming the government and making a mockery of the nation's leaders. Chechnya needs men like Kadyrov.[23]

The support obtained by the Kadyrov's in Chechnya demonstrates not only that top-down approaches to legitimacy-building can produce substantial results but also that a long-suffering population is inclined to accept as legitimate any authority, even if coercively imposed, capable of guaranteeing the provision of law and order. Although interviewees were well aware that political processes in Chechnya "mean nothing"[24] and that elections are "a pure formality,"[25] they nevertheless considered the imposition of a *de facto* dictator as the price to pay for enjoying the return to peaceful life after years of all-out warfare.

More recently, the Kenyan government adopted a similar approach to legitimacy-building in the Somali southern territory of Jubaland, where the presence of the Islamist insurgent group known as Al-Shabaab threatens the country's external and internal security. To contain the spillover of insurgent violence from the Juba Valley into Kenya's borders, the Kenyan government in 2010 started promoting the creation of a buffer zone between its national territory and the areas controlled by Al-Shabaab (International Crisis Group 2012: 2). As the project of creating a local government rested on the endorsement of a capable leader, the Kenyan government tasked the former Somali Minister of Defence, Mohamed Abdi Mohamed, to identify a suitable candidate. The choice fell on

Ahmed Madobe, a former Jihadi insurgent warlord, who entered in conflict with his one-time comrades-in-arms (Bruton & Williams 2014: 107). As explained by Mohamed, Madobe was an ideal candidate for the task ahead:

> The Kenyan army was prepared to help us. I was in charge of raising Somali troops. We needed a leader, one with extensive military experience and good knowledge of the enemy. I immediately thought of Madobe, as he seemed sincere in his repentance.
>
> (Meyerfeld 2019)

Accepting the offer, Madobe swiftly seized the upper hand in the contest with other potential candidates, self-appointing himself in 2013 as the "elected president" of Jubaland at an assembly of around 500 local tribal leaders (Moe 2017: 129). Head of a 5,000-strong security apparatus, Madobe has spearheaded the fight against Al-Shabaab in the region and contributed to reduce the numbers of insurgent fighters crossing the border into Kenya (Felbab-Brown 2020: 132). Madobe's ability to contain the insurgency and establish a situation of relative stability in Jubaland helped consolidate his power position and gain legitimacy from both international and domestic circles. While Madobe received an official endorsement from the Special Representative of the United Nations (U.N.) Secretary-General Nicholas Gray at his administration's inauguration in 2014 (U.N. 2014), a former U.N. staff member in Somalia reconfirmed in 2019 that Madobe enjoys the support of local key actors: "Madobe has shown strong campaigning, has got endorsement from key local clans and shown resilience to stand against [President Mohamed] Farmaajo's Federal government power and resources to unseat him" (Mutambo 2019). By endorsing a popular authority figure as Jubaland's leader, the Kenyan government contained the infiltration of Jihadi insurgents from the Somali border and stabilised a region where intra-tribal hostiles can easily escalate into open warfare. In Jubaland, the co-optation of a local leader, not the bottom-up participation of the local tribes to the country's political life, has best served the Kenyans' anti-insurgent containment strategy.

Third pillar: economic development

At the beginning of 2000, Chechnya was dangerously close to resembling a war-torn wasteland. With its "industrial base, social infrastructure, public and private housing, transport links, and engineering capabilities almost completely destroyed" by artillery shelling and carpet bombings, the Chechen population was enduring dreadful economic conditions, with many individuals forced to join rebel groups and crime syndicates to provide for their families (Basnukaev 2014: 76; Galeotti 2002). For the purposes of revitalising the economy and restoring an appearance of normal life, the Russian government granted Chechnya an unprecedented influx of federal subsidies to rebuild hospitals, schools, roads, households, and industrial complexes. Between 2000 and 2010, Moscow transferred from federal to Chechen coffers the impressive amount of 30 billion dollars—money that the Chechen leaders used at their own discretion to stimulate Chechnya's economic recovery (Alexseev 2011).

Following what Matveeva defined as an "essentially Soviet approach" to economic development (Matveeva 2007: 6), the Kadyrov clan prioritised the country's reconstruction along two main lines of action. The first step taken to stabilise Chechnya was the recovery of the country's public facilities and urban settlements destroyed during the war. At the end of 2009, Chechnya appeared as a regenerated nation. While entire neighbourhoods were being brought back to their pre-war status, medical and educational facilities were restored to "the same level as before the wars" (LandInfo 2012: 13). Although the reconstruction of householdings and critical infrastructures occurred all across Chechnya, it was in Grozny, Kadyrov's "most impressive gift to the Chechen nation," in which Moscow's subsidies have been put at their best use (Erbslöh 2016: 208). Flattened to the ground during the conflict, nowadays Grozny looks like a world-class capital, with luxurious buildings, fancy cafes, and the recently inaugurated "Europe's largest mosque" standing as a symbol of Chechnya's economic rebirth (Dutton 2019). If Grozny's transition from devastated battleground to modern city impressed the international audience, the effects produced on the Chechen population were even greater. While a Chechen economist confessed that "what has been achieved [in Grozny] is . . . most astonishing," the comment of one interviewee confirmed the strategic significance of the city's reconstruction: "when I visit Grozny, I cannot even remember how the city looked like during the war, everything is brand-new. There is nothing that reminds people of those tragic years"[26] (Hille 2015). By hiding the scars of war behind a façade of economic lavishness, Kadyrov demolished the insurgents' popular appeal, signalling to the population that defying the authority would have risked jeopardising the prosperity obtained after years of laborious progress. While Grozny's reconstruction fulfilled important propagandistic purposes, the organisation of Chechnya's national economy permitted the regime to minimise the risk of a resurgence in rebel activity.

Preventing individuals from finding profitable employment outside of state-controlled enterprises constituted the strategic keystone of Kadyrov's development agenda. Although the regime encourages the proliferation of small private businesses, such as mini-markets, cafes, and restaurants, as a way to consolidate the illusion of a thriving economy, enterprises that generate significant revenues are precluded from operating without the government's direct approval. As underscored by one interviewee maintaining regular contact with people living in Chechnya,

> private businesses are always controlled by governmental organisations, there is no way to avoid it. For example, anyone who wants to open a business must rent a property, but all properties are owned by the state, and the state alone decides who can and can't open a business. Usually, only those loyal to Kadyrov are allowed to operate remunerative activities."[27]

The government's pervasive presence in the country's economic life was evident to another interviewee, who concluded that, in Chechnya, "you do not own your own business, as the authorities are in charge of everything. If the government does not like your business proposal, there is no chance for you to start that activity."[28] By keeping the private market in a state of chronic underdevelopment and by offering people few "exceedingly" underpaid jobs, the regime forces entire families to rely on the economic inducements provided by the government as their only viable source of income (Halbach 2018: 25). In doing so, Kadyrov not only forces people into providing allegiance to the pro-Russian regime but also limits the insurgents' possibility of extracting enough resources from the population to sustain a protracted campaign of partisan warfare.

Despite wages being kept at a low level to deprive potential rebels of active popular support, the regime prevents impoverished people from joining insurgent groups out of economic necessity by providing them with the means necessary to conduct a decorous life. While an interviewee confirmed that the government takes care of the poor and makes sure that "no one starves,"[29] another explained that "Kadyrov builds houses, puts furniture inside, and gives the keys to the homeless as a way to win the hearts of the population."[30] Kadyrov's apparent generosity towards those in need should not conceal the strategic objectives accomplished by the "charitable" activities sponsored by the government. In Chechnya, the population knows that economic inducements constitute a reward for loyalty and that suspected insurgent sympathisers are systematically sentenced to a life of hardship for the crime of disobeying the regime. Acting as a form of collective punishment, the government's practice of denying food, shelter, healthcare, and education to people associated with the rebellion produces a strong deterrent against potential non-compliers. "You must remember that, in Chechnya, the government is the only job provider, the only entity that decides who gets a job and for how long," stressed an interviewee, who further specified that:

> if someone is caught disrespecting the authority, the government precludes that person from finding employment ever again, condemning his family to a life of misery. By controlling the job market, the regime controls the population and dissuades people from challenging its rule.[31]

In other terms, Kadyrov gives the population a choice: people either can enjoy a comfortable life under the government's rule or can choose to support the rebels at the risk of being caught and sentenced to live a life of starvation and disease. Given the alternatives, it is not surprising that the vast majority of the population chose the former over the latter.

While the weaponisation of economic inducements facilitated the task of discouraging anti-incumbent activities, Chechnya's reconstruction has been carried out under "the watchful eye" of the Russian government, which prevents Grozny from seizing the wealth necessary to decrease its dependency on federal subsidies (Souleimanov & Jasutis 2016: 122). The hierarchical power dynamics existing between Moscow and Grozny are particularly visible in the economic activities related to the extraction of oil—a resource that could provide the regime with a steady influx of considerable revenues. While the Russian government does not allow exploration and drilling activities to take place without its direct approval, all taxes on oil revenues generated in Chechnya are to be paid exclusively to federal institutions (Bodner 2015). By preventing the Chechen leadership from accessing the resources necessary to boost profitable sectors of the local economy, Moscow forces Grozny to rely for more than 80% of its annual budget on federal subsidies, precluding the country from drifting away from the federal centre (Fuller 2017). Instead of encouraging the emergence of a private sector as advocated by democratic COIN guidelines, the Russians left no choice to the population but submitting to the government in return for economic gains. By tightly controlling Chechnya's economy, Moscow reduced the insurgents' potential for collective action whilst preventing the co-opted elite from "going rogue" and attempting to challenge Russia's supremacy.

This coercion-centred approach to economic development has been extensively implemented by the regime of President Bashar al-Assad in the ongoing Syrian Civil War. In an effort to showcase strength to the Syrian population, the regime has engaged in the wanton destruction of public infrastructures, including schools, hospitals, and power plants, as a way to depopulate insurgent-controlled areas and deprive the rebels of crucial popular support. Bakeries situated in territories outside government-controlled areas have been systematically and repeatedly targeted by airstrikes ever since the outbreak of conflict in 2011 (Arnold 2013). The regime's starvation policy aimed at simultaneously accomplishing two objectives, namely making life under the rebels impossible and ensuring that food supplies never run out in government-held territories. While the deliberate destruction of bakeries in rebel-held territories helps the government in preventing the population from providing active support to the insurgents, the government does its best "to maintain the bread subsidy in areas it controls by ensuring that bakeries are open, well stocked with flour, and consistently distributing the foodstuff" (Martínez & Eng 2017: 135). Given that choosing to remain in rebel-held areas might signify a slow death by starvation, it should come as no surprise that entire communities have moved away from rebel-held areas to pro-government territories. Hospitals are another preferred target of the regime's airstrike campaigns. "When I am in the hospital, I feel like I am sitting on a bomb," stated a doctor from a hospital located in the last rebel-held territory in the Idlib province, stressing the fact that the attacks are by no means incidental: "I wish I could say that targeting a hospital in Syria is unique, but it is not. The field hospital I direct in Sarmin has been targeted and hit by airstrikes more than a dozen times" (SAMS 2015). By systematically striking healthcare facilities and killing the medical personnel working in rebel-held areas, the regime has deprived the population of a crucial means of survival, thus forcing entire communities to abandon their homes and transfer to government-held territories. The profound psychological and physical damages inflicted by these airstrikes on the population living under the rebels' control emerge from the accounts of people who witnessed first-hand the attacks:

> After the attack [at the hospital], electricity generators stopped, bakeries shut down. People started leaving with their belongings. Even roaming merchants, vegetable sellers, the shops—they all closed, and all the people left. Whoever was not scared and did not leave because of the strikes eventually left because of the lack of basic services.
>
> (Human Rights Watch 2020: 41)

In government-controlled areas, the regime has leveraged its undiscussed control over the country's economic resources to punish dissenters and reward supporters. As explained by Asseburg, the regime's reconstruction programmes were designed to both consolidate patronage networks of regime supporters and penalise suspected dissenters: "population groups that are regarded as unreliable experience collective punishment and displacement, especially in politically and strategically important areas" (Asseburg 2020: 8). In Damascus, Syria's capital, the regime has systematically seized the properties of people accused of purporting terrorist activities, enabling trusted businessmen to raze them to the ground and make space for luxury residential buildings instead (Daher 2019: 3). According to critics of Syria's reconstruction schemes, the government has been "pushing out impoverished communities seen as centers of opposition support and replacing them with wealthier ones more likely to be loyalists" (Karam 2018). In other cases, the regime's intelligence services run background checks on people wishing to return to their homes, denying access to families suspected of sharing ties with the rebellion: "some people will get an approval. Others won't . . . Hopefully, our name gets on the list and they let us back" (Human Rights Watch 2018). Even humanitarian projects financed by international non-governmental institutions have been weaponised by the regime, which diverted aid and relief programmes away from anti-incumbent areas to punish dissenters and reward loyalists. As explained by a local activist, there is little chance of seeing projects being approved in areas considered as potentially sympathetic to the rebels:

> In Syria, you barter with the government for projects, everyone knows this. As a humanitarian, I say I will rehabilitate schools in the area. The government comes back and says how about these areas instead? Back and forth, until I commit to their areas to get approval for my projects.
>
> (Human Rights Watch 2019: 23)

By turning economic development into a weapon to be wielded against potential dissenters, the Syrian regime has cemented a large network of loyal supporters and effectively signalled that non-compliance will not be tolerated in post-conflict Syria.

Brutalisation and success in Chechnya

In October 2008, less than a year before Russia's then-President Dmitry Medvedev announced the end of the COIN operations in Chechnya, American journalist Steele (2008) suggested that Moscow's victory would have been complete and definitive: "like it or not, Russia has won this war." If this assessment seemed premature to some in 2008, the past 10 years of Chechen history have not disproven Steele's evaluation. In the fight against the insurgents, Russia proficiently met all the criteria identified by this book as indicators of full-fledged mission success.

Ever since the formal end of the Second Russian-Chechen War, the frequency and lethality of insurgent-related attacks registered in Chechnya displayed an exponentially downward trend. According to one of the most reliable datasets on battle fatalities in Chechnya, the 2010–2019 period saw a 93% overall decrease in the number of insurgency-related deaths, with the majority of casualties being militants killed during armed clashes with the security forces (Caucasian Knot 2012, 2018, 2019a, 2019b, 2019c). With the year 2017 standing as an exception due to groups of avengers staging indiscriminate attacks in retaliation for the COIN operations performed in previous years, Figure 4.1 shows a substantial decrease in the levels of insurgent activity registered in Chechnya, a trend that reached the historical minimum of only six reported casualties in 2019 (Souleimanov 2017b). Although the drop in insurgent activity can be partially attributed to the outflow of fighters to foreign hotbeds of insurgent warfare (Aliyev 2015), the sharp decline in the number of attacks that occurred in Chechnya can hardly be explained without referring to the results obtained throughout years of intensive COIN operations. The almost complete collapse of the insurgency' support infrastructure is another clear indicator of Russia's success. As demonstrated by Souleimanov and Aliyev in several publications discussing the decline of pro-insurgent popular support in

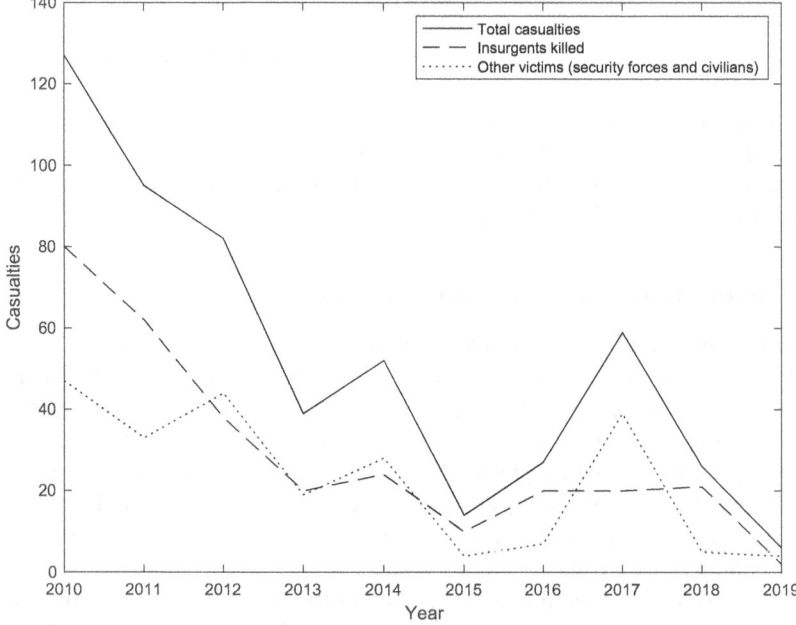

Figure 4.1 Insurgent-related casualties registered in Chechnya (2010–2019)

the region (Souleimanov & Aliyev 2015a, 2017), the unbearable pressure exercised by the regime on the population to stop supporting the insurgents became so strong as to compel entire villages to deny shelter and protection to the rebels.

These military-centred metrics of success complement the set of criteria associated with population-centric standards of evaluation. In Chechnya, the Russians not only kept the co-opted government in charge throughout the entirety of the armed confrontations but also reduced to a minimum the risk of secession by endorsing an indigenous ally fully loyal to the Russian leadership. As Kadyrov himself regularly affirms, he is proud to be "Putin's man" and to be "ready to die for him, to fulfil any order" (The Jamestown Foundation 2007; Osborn & Solovyov 2017). The unconditioned loyalty of the Chechen dictator to the Russian President allowed the Kremlin to tame the insurgency without granting any concession to its leadership. As declared by Putin, "Russia does not negotiate with terrorists, it destroys them" (Kremlin 2004). By strictly following this agenda up until the end of the COIN struggle, Moscow consolidated its uncontested supremacy over the rebellion-prone republic and minimised the potential for an insurgent's comeback in the long term. These results indicate that the hearts and minds approach is far from constituting a universal recipe for victory, as brutalisation-centred measures can be equally effective against grassroots insurgencies.

Although Russia's success in Chechnya calls for a re-evaluation of several techniques categorised by the dominant literature as "bad COIN practices," scholars researching the long-term effects of Moscow's strategy underscored that the brutalisation paradigm is not exempt from potential backlashes. If Chechenising the conflict allowed the Russians to divide and conquer the splintered factions of the Chechen resistance, delegating powers to an indigenous ally disposing of a 7,000-strong personal militia might engender potential repercussions for Russia's national security. As noted by Souleimanov, Abbasov, and Siroky, Putin's strategy for Chechnya presents several cracks that could lead the "frozen" conflict towards a "violent thawing period" (Souleimanov et al. 2019: 90). First, Moscow cannot predict with certainty whether Kadyrov will continue to remain loyal to its Russian masters. Given the imperative of preserving the integrity of Russia's southern frontiers, Moscow has few available options other than sustaining the costs necessary for securing the allegiance of its Chechen vassal, with the consequence that Kadyrov is becoming increasingly bolder in its requests to the Russian Federation. As Politkovskaya famously commented, "a little dragon has been raised by the Kremlin. Now they need to feed it. Otherwise, it will spit fire" (quoted in Knight 2017: 141). Second, Kadyrov's repressive methods broke the insurgency's backbone, but only at the cost of antagonising hundreds of individuals who are now waiting for propitious times to commence violence. As one former fighter put it:

> there are a lot of people currently opposing the regime, but they cannot say or do anything [due to the climate of repression]. When the president retires, I am certain that many will take blood revenge upon government officials, the *Kadyrovtsy*, and their civilian collaborators.[32]

While the cracks in the foundations of Russia's strategy foretell the risk of a conflict escalation, the evidence submitted in the present study scales down the assessment of Chechnya's status quo as "untenable in the *longue durée*" (Souleimanov et al. 2019: 100). A first indicator suggesting a stronger resilience of Chechnya's political situation emerges from the last two decades of incessant work carried out by the Kremlin to permeate every aspect of the country's social, political, economic, and cultural life. Far from considering the co-optation of the Kadyrov clan as a concluding step in the pacification effort, Moscow consistently makes sure that Chechnya will never again be able to pursue a secessionist agenda. As one interviewee remarked, this restless activity brought substantial benefits to Moscow: "without Russia, I don't think that Chechnya could self-sustain itself. The Russians seized all the resources that Chechnya needs to be functionally independent. There is no doubt that Russia achieved its goals."[33] Should a premature change in Chechnya's leadership take place, many Chechens are aware that Moscow would immediately step in to de-escalate a potential crisis. While a deterioration in the relationship between Putin and Kadyrov could create temporary instability, it would be surprising seeing Moscow caught unprepared to react to such a predictable eventuality. As one interviewee stated:

> I don't see how this [scenario] would drastically change the situation: the Russian Federation has deep roots in our society and controls every aspects of the Chechen political sphere . . . the situation will not change after Kadyrov retires; Moscow will identify his successor, who will do exactly what the Russian Federation expects the ruler of Chechnya to do. I don't see war emerging once Kadyrov will no longer be president.[34]

The tight control exercised by the Kremlin over Chechnya's political institutions is complemented and reinforced by the second indicator of regime resilience related to the demographic composition of the Chechen population. As estimated by several human rights organisations (World Peace Foundation 2015; RFE 2005), the death toll of approximately 160,000 civilians killed throughout the wars weighed enormously over a nation of around 1 million people, which was further reduced by the outflow of thousands of refugees fleeing to Europe. Occurred in less than 20 years, the depletion of the adult population radically changed Chechnya's age structure, and with it the socio-cultural composition of its social texture. As explained by one interviewee:

> There is a sharp difference between pre and post-war Chechens. In today's Chechnya, you don't see around many people of my age,[35] as many died during the wars. Those who got killed were the brave ones, those willing to sacrifice for their country . . . The old generation is no longer capable of revitalising the rebellion. It is up to the new generation now, and the beliefs, aspirations, education, and socio-cultural roots of the youth will be decisive in determining the likelihood of a new conflict. But the legacy of the two wars is fading away, and the new generation is far different from the previous one—it does not want to fight the government. I am convinced that even in front of a political crisis, such as a change in the Chechen leadership, the young generation would not pick up a rifle to fight for independence.[36]

With a population largely comprising people raised and educated under the banners of the new pro-Russian regime, Kadyrov—and Putin in turn—ensured that the dream of independence would fall out of fashion among a generation of Chechens that, aware of what their families endured during the wars, no longer desires to "spill blood" in the pursuit of their fathers' ambitions.[37] While this evidence does not refute the fact that Moscow's political edifice for Chechnya has been built on unstable foundations, the measures taken to contain potential fallouts suggest that the current status quo might be more resilient than what is generally assumed. The attested success of the brutalisation paradigm in Chechnya holds significant implications for Western counterinsurgents.

Implications for Western counterinsurgents

Sharing the perspective of many Western analysts, military historian Miakinkov concluded in his study of the Chechen Wars that "there is little indeed" a democratic state can absorb from the experience of counterinsurgents following the brutalisation paradigm (Miakinkov 2011: 674). Miakinkov's assessment bears some truth, as wantonly brutal tactics remain largely inaccessible to liberal democracies—and rightly so. Arguing in favour of a pivot to brutalisation-centred measures would break any conceivable ethical and moral boundary necessary to preserve the dignity and integrity of human life, especially in the chaotic context of civil war and insurgency. Through an in-depth scrutiny of brutalisation-centred measures, this book shows that victory in COIN warfare is often achieved in bloody and disdainful ways. Yet, its findings demonstrate that these measures *are* effective, and it is exactly because of their effectiveness that they continue to find widespread application in present theatres of COIN warfare. Shedding light on the mechanisms of the brutalisation paradigm, this book signals that fighting an insurgency on the ground might look substantially different from the polished accounts advanced in Western military manuals. It is important to explore alternative pathways to COIN warfare because *a priori* refusing to extrapolate useful lessons from the experience of successful counterinsurgents only slows down the progress towards better-refined COIN techniques. Fighting insurgency is an extremely challenging endeavour, and as long as decision-makers lack a deeper understanding of which techniques do and do not work in a given operational battleground, they will be prevented from optimising their strategic approaches. We move away from mythologised interpretations of COIN warfare, without however losing focus on the ethical and moral imperatives of fighting a war against an elusive enemy hiding amongst innocent civilians. In an effort to overcome the psychological barrier placed against the study of brutalisation-centred measures, this book distils from the Chechen case, a list of takeaways meant to improve the strategic blueprints available to Western counterinsurgents.

Intelligence base: play the indigenous card

The deployment of indigenous forces for intelligence-gathering purposes constituted the trump card of Moscow's COIN efforts during the Second Russian-Chechen War. Marking a watershed in the process of identification and neutralisation of enemy fighters and civilian supporters, the provision of actionable intelligence by defected insurgents and their relatives effectively decapitated the insurgency's leadership and facilitated the disruption of the rebels' organisational infrastructure. Although this book discourages Western practitioners from resorting to torture and other illegal intelligence collection techniques, the Chechen case study demonstrates that setting up a capillary network of native collaborators magnifies the amount and quality of collected HUMINT. In the light of these findings, practitioners would find profitable, employing more invasive intelligence techniques geared towards the exploitation of indigenous sources. In particular, Western strategists could find it helpful to learn from the experience of the Phoenix

Program (1965–1972), a Vietnam-era intelligence effort aimed at neutralising the insurgency's political and administrative infrastructure, known as the Viet Cong Infrastructure (VCI). To obtain the information necessary to locate, capture, and/or neutralise suspected VCI cadres, Phoenix relied on "a nation-wide" network of local informants capable of identifying and locating VCI members (Clair & Cockburn 2001). Capable of penetrating deep into every hamlet, Phoenix officers obtained constant streams of timely and actionable intelligence *while* abiding by strict legal and operational requirements (MACV 1970a). The high degree of effectiveness achieved by the Phoenix Program finds confirmation in captured enemy documents, in which the VCI leadership refers to Phoenix as "the most dangerous enemies of the Revolution" and instructs the guerrillas to employ all means necessary to "crash the head of the venomous snake Phoenix" (Finlayson 2015: 2; MACV 1970b: 6). By establishing a permeating surveillance system within the host society, counterinsurgents can better triangulate the information collected from trusted informants without compromising the legal requirements for intelligence operations conducted according to the hearts and minds approach.

IO base: pound the enemy

Russia's all-out media offensive in Chechnya incorporated the typical features of Soviet-era propaganda campaigns: aggressive in its nature, pounding in its execution, and merciless towards its target. Quick in silencing the enemy's propaganda, the Russians achieved impressive results by bombarding the population with contents meant to dehumanise the rebels and deepen the cleavages already existing within the Chechen society. Conversely, Western counterinsurgents have been struggling to outsmart insurgent groups in the information domain of COIN warfare. While the U.S. information campaign in Iraq "lack[ed] resonance and relevance among ordinary Iraqis" (Garfield 2007: 27), the Coalition forces operating in Afghanistan largely failed to discredit the Taliban's propaganda, ultimately signalling "more vulnerability than strength" to the local population (Rahmani & Lawrence 2018). It follows that Western IO officers must be ready to confront the insurgents at their own game, communicating decisiveness and resolve to their target audience with all means at their disposal. This will require discarding the "political correctness" characterising democratic approaches in favour of more aggressive techniques aimed at stigmatising the insurgents and curtailing their popular appeal. The benefits of abiding by more assertive IOs have been noticed by several officers deployed in recent theatres of COIN warfare. In recalling his experience as an IO officer in Iraq, Major

General Molan explained that a prompt reaction to enemy propaganda was an essential prerogative of his work:

> To prepare for propaganda attacks from a hostile media, General [George W.] Casey gave me one very important task. He directed me in no uncertain terms to ensure that there was no more than a one-hour turnaround between an allegation appearing in the media and our response being fired back.
>
> (Molan 2009: 50)

Similarly, Lieutenant General Metz stresses the importance of seizing the initiative in the propaganda war:

> I am absolutely convinced that we must approach IO in a different way and turn it from a passive warfighting discipline to a very active one. We must learn to employ aggressive IO. We cannot leave this domain to the enemy; we must fight him on this battlefield and defeat him there.
>
> (Metz et al. 2006: 4)

This was further confirmed by Army Captain Flor, who argues in an article for *Military Review* that waging a more decisive IO warfare decreases the insurgents' ability to spoil the successes achieved by the COIN force:

> our task force set out to ensure that we would not have another defeat snatched from the jaws of victory due to a lack of aggressive information operations. For the next year, our information operations became tactical-unit battle drills that we executed with vigor . . . The result was a more coherent and effective counterinsurgency effort.
>
> (Flor 2010: 58)

The benefits of waging a more aggressive IO campaign are made even more conspicuous if one considers that IHL precepts do not prohibit states from utilising their superior media apparatuses to force the insurgency's propaganda machine into silence. As explained by Schmitt, the use of stealth, misinformation, and other ruses of warfare is allowed under the LOAC for as long as the target does not have an affirmative reason to trust the attacker: "[Lawful] ruses are planned to mislead the enemy, for example, by causing him to become reckless or choose a particular course of action" (Schmitt 1992: 617; Hague Convention IV 1907: art. 24). Hence, counter-insurgents are legitimately entitled to take all necessary measures to expose the insurgents' wrongdoings and respond in kind to the enemy's deceitful propaganda. While LOAC guidelines prevent IO officers from harassing the population with brainwashing and/or deceiving messages, the recourse to assertive instruments of counterpropaganda warfare is deemed legally permissible when the intended targets are the members of the insurgency's network (Wingfield 2009). By saturating the area of operations with assertive counterpropaganda messages, Western IO officers would more effectively seize and maintain the information initiative against the insurgents and their political apparatus.

Use of force: tougher does not mean barbaric

As long as the population actively supports the rebels, COIN endeavours are destined to fail. Learned the hard way during the First Russian-Chechen War, the marginalisation of the insurgents from their civilian sympathisers constituted the top-priority objective of Russia's second round of COIN operations. While Moscow's use of indiscriminate violence was patently criminal and deserves condemnation, its effectiveness signals that a COIN campaign can hardly be successful if the population is left free to decide where its allegiance resides, as all odds suggest that brothers, husbands, and sons will always be chosen over the government. This is valid for the brutalisation paradigm as well as for the hearts and minds approach. After all, as argued by Newsinger, defeating insurgency without recurring to a significant dose of coercion *is not* what hearts and minds are all about:

> hearts and minds campaigns are not soft-hearted exercises in sentimentality carried out by social workers in uniform. Alongside the reforms and concessions, the material advantages that are intended to win over the "hearts" of the local population, there is the use of force to focus their "minds."
>
> (Newsinger 2002: 145)

In Afghanistan, coercion has often accompanied the more kind-hearted operations that the U.S. forces perform in favour of the local population—and often scoring noticeable successes. This was the case of the destruction of entire villages and the resettlement of their inhabitants in built-anew settlements in the province of Kandahar. Because the Taliban enjoyed widespread support amongst the local population, the U.S. forces were suffering heavy casualties in vain attempts to drive the insurgents out of the area (Belcher 2018: 100). When less invasive means to clear the insurgents out were proven ineffective, the commanding officer Lieutenant Colonel Flynn opted for a more coercive approach that would have granted success without causing excessive harm to the population. Although the decision to raze entire villages was not taken light-heartedly, the commitment of granting a comfortable life to the families deprived of their homes was taken to completion:

> If we didn't care about the local population, I would have thrown money at them and bid them farewell. We are committed to their future, and as far as I can tell they are walking side by side with us.
>
> (Flynn 2011)

A more decisive use of force cannot however fail to comply with the LOAC regulating the conduct of COIN operations. As stated by an Italian officer who served in Afghanistan, commanders who authorise kinetic operations without concerns for the local population's safety commit a breach of the legal requirements that play a fundamental role in hearts and minds operations: "no military opportunity could justify disregard for the lives of civilians" (Sperotto 2009: 21). According to IHL precepts, any attack that "may be expected to cause incidental loss of civilian life, injury to civilians, damage to civilian objects, or a combination thereof, which would be excessive in relation to the concrete and direct military advantage anticipated" is unlawful and therefore prohibited under the legal frameworks in use amongst Western militaries (ICRC 1977, art. 51, 5[b]). To satisfy these requirements, counterinsurgents must perform a "balancing test" necessary to weight the military necessity of carrying out a given operation against the potential hazards faced by non-combatants:

> The principle of proportionality requires a commander to conduct a balancing test to determine if the expected incidental injury resulting from an attack, including harm to civilians and damage to civilian objects, would be excessive in relation to the concrete military advantage anticipated to be gained from the attack.
>
> (U.S. Gov. 2017: 5-4)

When the use of coercive measures is calibrated according to IHL principles, Western counterinsurgents can resort to a more consistent and therefore effective use of kinetic force that is both strategically sound and legally permissible.

An important application of this takeaway refers to the use of drone strikes in the context of COIN warfare. While the technology in use on drones magnifies the pilot's ability to discriminate between insurgent fighters and innocent civilians, the often poor consideration given to the need for a balancing test can lead to unlawful *and* strategically counterproductive abuses of this otherwise important tool of warfare (Strawser 2010). When LOAC principles are disregarded, the indiscriminate use of kinetic strikes can prompt the kinsmen of individuals unjustly killed by the security forces into pro-insurgent violent mobilisation. As explained by Gurcan, "air bombardments [and drone strikes] by U.S. Forces in Afghanistan, which kill many innocent women and children, cause insurmountable agony and thirst for retaliation from the local population" (Gurcan 2016: 57). These dynamics are typified by an episode that took place in a Pakistani village in 2011. After that a U.S. drone strike accidentally killed 44 innocent local villagers, their kinsmen did not hesitate to rise up against the aggressor: "Americans don't spare us—not our children, nor our elders, nor our younger . . . This is why we have decided we will take blood revenge however we can" (Walsh & Habib 2011). It is thus imperative to carefully consider the legal aspects of COIN warfare not only because it is the right thing to do but also because failure to do so might compromise the attainment of important strategic objectives. By harmonising the principle of "minimum necessary force" with the calibrated use of coercive measures according to LOAC principles, a counterinsurgent can fully exploit the potential of its strategic toolkit without infringing the ethical and moral requirements necessary for the conduct of effective hearts and minds operation.

Political legitimacy: rely on local agents

The co-optation of the Kadyrov clan constituted the strategic mainstay of Russia's political scheme for Chechnya. While turning notorious foes into loyal friends facilitated the decapitation of the insurgency's leadership, endorsing charismatic figures as the new regime's leaders allowed the Russians to garner genuine popular support for the pro-Russian camp. The effectiveness of elite co-optation in COIN warfare finds application in both brutalisation and hearts and minds endeavours. Drawing upon his experience as a French officer in Algeria, Galula argues that relying on local agents can lead to success in legitimacy-building endeavours:

> It would be a mistake to believe that a counterinsurgent cannot get the population's support unless he concedes political reforms. However unpopular he may be, if he is sufficiently strong-willed and powerful, if he can rely on a small but active core of supporters who remain loyal to him because they would lose everything including their lives if the insurgent wins, he can maintain himself in power.
>
> (Galula 2006[1964]: 71)

This was confirmed by another veteran of the Algerian conflict, Roger Trinquer, who explained in his book on *Modern Warfare* that co-opting local tribal leaders constituted a key aspect of the COIN operations performed against the FLN insurgency: "A careful search of the population is necessary to find men capable of being leaders . . . The bulk of the population is by habit or tradition normally devoted to established authority and the forces of order" (Trinquer 1985: 33). As people living during periods of social disarray are prone to accept as legitimate any authority capable of restoring a resemblance of normal life, Western counterinsurgents would find more effective endorsing an authority figure *instead* of promoting electoral processes.

Relying on local agents has been effective in Oman, where the British confronted a determined insurgency rooted in the province of Dhofar (1963–1976). Aware of the population's animosity towards the local Sultan Said bin Taimyr, the British helped organise a *coup d'état* in favour of his son, Qaboos. More condescending than his father, Qaboos was willing to launch civil reforms meant to appease the population and boost the incumbent's control over society (DeVore 2012). As reported by a local British officer, the *coup* "was received with predictable jubilation" across Oman, and especially in the capital Muscat, where Qaboos's entry into the city "became an instantly mythologized moment in Omani history" (Takriti 2013: 200). By placing popular elites in leadership positions, counterinsurgents can postpone electoral processes to more peaceful times, hence preventing the insurgents from spoiling political processes and discrediting the incumbent as incapable of enforcing law and order across the territory.

Economic development: weaponise carrots

In Chechnya, people can live a decorous life provided that they refrain from engaging in behaviours sanctioned by the authority. Strong of its ability to identify and punish non-compliers, the Chechen government makes sure that only those individuals who submit to its rule obtain access to essential services and sources of income. By depriving insurgent sympathisers of the means necessary to provide for their families, the regime deters the rest of the population from collaborating with the opponent, raising the costs of defying the authority well above any potential benefit associated with supporting the rebellion. While this book strongly condemns the cruelty with which the Russians implemented their approach to economic development, the weaponisation of economic carrots carried out in a lighter configuration can be compatible with the hearts and minds approach. The important role that economic sticks play in the hearts and minds COIN campaigns clearly emerges from the analysis of the British COIN experience. In Malaya, for the first time in history, entire villages obtained access to British-provided public services, such as schools, medical aid stations, and community centres, but these benefits came with a price. As explained by Owens, hearts and minds were only one side of the British COIN:

> Collaboration was obtained because the British military and its proxies were in a position to deprive people of basic life necessities through physical coercion. Unless civilians cooperated by providing intelligence, they faced food rations, extended curfews, torture and even death.
>
> (Owens 2015: 187)

This experience was replicated in Kenya against the Mau insurgency (1952–1960). In an effort to deter the population from providing support to the insurgents, the British exploited economic resources to reward compliance and punish disobedience:

> Land reforms were introduced with the aim of developing a conservative elite of loyalist landowners, while simultaneously punishing their Mau Mau-supporting neighbours . . . Poorer loyalists who had no land were rewarded with access to government and other jobs, which were also denied to supporters of the insurgency.
>
> (French 2011: 195)

While these examples taken from the colonial history of COIN warfare show that coercive measures played a key role in early hearts and minds endeavours, nowadays' counterinsurgents are strictly prohibited from engaging in any form of collective punishment against the civilian population. As stated by article 33 and article 53 of the Fourth Geneva Convention of 1949, "reprisals against protected persons and their property are prohibited" and "any destruction by the Occupying Power of real or personal property belonging individually or collectively to private persons . . . is prohibited, except where such destruction is rendered absolutely necessary by military operations" (ICRC 1949). The practice of carrying out collective punishments not only violates fundamental IHL principles but it is also incompatible with the typology of COIN operations performed by Western counterinsurgents. In the aftermath of the 2003 invasion of Iraq, for instance, the U.S. troops' malpractice of demolishing the houses of alleged insurgents raised allegations of suspected "violation[s] of international humanitarian law" and likely contributed to exacerbate the popular resentment against the Coalition troops (Amnesty International 2003).

It is within the boundaries of legality that economic aid should be used in a more assertive configuration. Evidence from recent theatres of COIN warfare shows that misused development aid has in some instances increased the levels of insurgent activity, fuelled popular distrust towards the authorities in others, and in many cases incidentally financed the insurgents' cause. Although development aid is a fundamental tool for stabilising war-ravaged societies, it is in the counterinsurgent's interest to make sure that its investments are well placed. In these terms, allowing the population to reap the benefits of economic development aids while refraining from punishing non-compliance only damages the incumbent's credibility and contributes to protract the hostilities. By threatening to deprive insurgent backers of their chances of living a comfortable life under the incumbent's rule, Western counterinsurgents would reduce the potential for anti-incumbent collective action and minimise the risk of seeing their efforts nullified by individuals persisting in acting against the restoration of law and order. This more assertive approach bears the potential to achieve better results without infringing IHL principles. As explained by Pictet, the lawful use of punishments is not prohibited under Article 33: "[the prohibition on collective punishments] does not refer to punishments inflicted under penal law, i.e., sentences pronounced by a court after due process of law, but penalties of any kind inflicted on persons or entire groups of persons, in defiance of the most elementary principles of humanity, for acts that these persons have not committed" (Pictet 1958: 225).

Notes

1 Interview with PG20207.
2 Interview with PG20202.
3 Interview with PG20208.
4 Interview with PG20207.
5 Interview with PG20205.
6 Interview with PG20203
7 Interview with PG20204.
8 Interview with PG20201.
9 Interview with PG20202.
10 Interview with PG20204.
11 Interview with PG20208.
12 Interview with PG20203.
13 Interview with PG20205.
14 Interview with PG20205.
15 Interview with PG20204.
16 Interview with PG20208.
17 Interview with PG20203.
18 Interview with PG20202.
19 Interview with PG20204.
20 Interview with PG20202.
21 Interview with PG20204.
22 Interview with PG20203.
23 Interview with PG20206.

24 Interview with PG20204.
25 Interview with PG20202.
26 Interview with PG20202.
27 Interview with PG20203.
28 Interview with PG20202.
29 Interview with PG20207.
30 Interview with PG20206.
31 Interview with PG20206.
32 Interview with PG20208.
33 Interview with PG20207.
34 Interview with PG20202.
35 Approximately between 55 and 65 years old.
36 Interview with PG20203.
37 Interview with PG20207.

References

Alexseev, Mikhail. "Rubles Against the Insurgency: Paradoxes from the North Caucasus Countries." *PONARS Eurasia Policy Memo* 157, May 2011. www. ponarseurasia.org/memo/rubles-against-insurgency-paradoxes-north-caucasus-counties.

Aliyev, Huseyn. "Conflict-Related Violence Decreases in the North Caucasus as Fighters go to Syria." *The Central Asia-Caucasus Analyst*, 1st April 2015. Available at: www.cacianalyst.org/publications/analytical-articles/item/13171-conflict-related-violence-decreases-in-the-north-caucasus-as-fighters-go-to-syria.html.

Amnesty International. *Iraq: US House Demolitions Could be Illegal*, 21st November 2003. Available at: www.amnesty.org.uk/press-releases/iraq-us-house-demolitions-could-be-illegal.

Arnold, David. "Bombs Target Syrian Bakeries, Customers." *VOA News*, 29th January 2013. Available at: www.voanews.com/middle-east/bombs-target-syrian-bakeries-customers.

Asseburg, Muriel. "Reconstruction in Syria: Challenges and Policy Options for the EU and its Member States." *SWP Research Paper* 11, 2020.

Baev, Pavel K. "The Challenge of 'Small Wars' for the Russian Military." In *Russian Military Reform 1992–2002*, edited by Aldis, Anne C. & McDermott, Roger N. (London: Frank Cass. 2003). ISBN: 0-7146-5475-2.

Baldinetti, Anna. *The Origins of the Libyan Nation: Colonial Legacy, Exile and the Emergence of a New Nation-State* (London: Routledge, 2010).

Basnukaev, Musa. "Reconstruction in Chechnya: At the Intersection between Politics and Economy." In *Chechnya at War and Beyond*, edited by Le Huérou, Anne (Oxon: Routledge, 2014). ISBN: 978-0-415-74489-8.

Baumann, Robert F. "Russian-Soviet Unconventional Wars in the Caucasus." *Leavenworth Papers* 20, 1993.

Belcher, Oliver. "Anatomy of a Village Razing: Counterinsurgency, Violence, and Securing the Intimate in Afghanistan." *Political Geography* 62, 2018: 94–105. https://doi.org/10.1016/j.polgeo.2017.10.006.

Blank, Stephen. "Russian Information Warfare as Domestic Counterinsurgency." *American Foreign Policy Interests* 35, 2013: 31–44. https://doi.org/10.1080/108 03920.2013.757946.

Bodner, Matthew. "Kadyrov: Elders Say Chechnya Has as Much Oil as Saudi Arabia." *The Moscow Times*, 18th June 2015. Available at: www.themoscow times.com/2015/06/18/kadyrov-elders-say-chechnya-has-as-much-oil-as-saudi-arabia-a47493.

Bruton, Bronwyn E. & Williams, Paul D. *Counterinsurgency in Somalia: Lessons Learned from the African Union Mission in Somalia, 2007–2013* (MacDrill Air Force Base: Joint Special Operations University, 2014). ISBN: 978-1-933749-90-7.

Caucasian Knot. *Armed Incidents and Victims in Northern Caucasus*, 2018. Accessed 6th April 2020. Available at: www.eng.kavkaz-uzel.eu/articles/reduction_number_victims_2018/.

Caucasian Knot. *In Quarter 1 of 2019, in Chechnya, Three People Fell Victim to Armed Conflict*, 2019c. Accessed 29th April 2020. Available at: www.eng.kavkaz-uzel.eu/articles/48372/.

Caucasian Knot. *In Quarter 2 of 2019, in Chechnya, Three People Fell Victim to Armed Conflict*, 2019b. Accessed 29th April 2020. Available at: www.eng.kavkaz-uzel.eu/articles/48457/.

Caucasian Knot. *In Quarter 4 of 2019, in Chechnya, No One Fell Victim to Armed Conflict*, 2019a. Accessed 29th April 2020. Available at: www.eng.kavkaz-uzel.eu/articles/50077/.

Caucasian Knot. в ходе вооруженного конфликта на северном кавказе в 2011 году погибли и были ранены 1378 человек (*During the Armed Conflict in the North Caucasus in 2011, 1378 People Were Killed and Injured*), 2012. Accessed 6th April 2020. Available at: www.kavkaz-uzel.eu/articles/198756/?redirected=www.kavkaz-uzel.ru.

Chagnon, Napoleon. "Life Histories, Blood Revenge, and Warfare in a Tribal Population." *Science* 239(4843), 1988: 985–992. Available at: www.jstor.org/stable/1700080.

Clair, Jeffrey & Cockburn, Alexander. "Phoenix and the Anatomy of Terror." *CounterPunch*, 8th November 2001. Available at: www.counterpunch.org/2001/11/08/phoenix-and-the-anatomy-of-terror/.

Cohen, Ariel. *Russia's Counterinsurgency in North Caucasus: Performance and Consequences* (Carlisle Barracks: Strategic Studies Institute, U.S. Army War College, March 2014). ISBN: 1-58487-606-9.

Daher, Joseph. "The Paradox of Syria's Reconstruction." *Carnegie Middle East Center*, 2019.

Dannreuther, Roland & March, Luke. "Chechnya: Has Moscow Won?" *Survival* 50(4), 2008: 98. https://doi.org/10.1080/00396330802329030.

Del Boca, Angelo. *Italiani, Brava Gente? Un Mito Duro a Morire* (Vicenza: Neri Pozza Editore, 2005). ISBN: 978-88-545-0495-0.

DeVore, Marc R. "A More Complex and Conventional Victory: Revisiting the Dhofar Counterinsurgency, 1963–1975." *Small Wars & Insurgencies* 23(1), 2012: 144–173. https://doi.org/10.1080/09592318.2012.632861.

Dorronsoro, Gilles. *Revolution Unending: Afghanistan, 1979 To the Present* (London: Hurst & Company, 2005). ISBN: 1-85065-683-5.

Dutton, Jack. "Chechnya Inaugurates Europe's Largest Mosque." *The Nation*, 23rd August 2019. Available at: www.thenationalnews.com/world/europe/chechnya-inaugurates-europe-s-largest-mosque-1.901768.

Erbslöh, Gisela. "Seeking Chechen Identity between Repression and Self-Determination under the Ramzan Kadyrov Regime." *Region: Regional Studies of Russia, Eastern Europe, and Central Asia* 5(2), 2016: 201–224. https://doi.org/10.1353/reg.2016.0016.

Evans-Pritchard, E. E. *The Sanusi of Cyrenaica* (Glasgow: Oxford University Press, 2008[1949]).

Fasanotti, Federica S. *'Vincere!' The Italian Royal Army's Counterinsurgency Operations in Africa, 1922–1940* (Annapolis: Naval Institute Press, 2020). ISBN: 9781682474280.

Felbab-Brown, Vanda. "The Problem with Militias in Somalia." In *Hybrid Conflict, Hybrid Peace: How Militias and Paramilitary Groups Shape Post-Conflict Transitions*, edited by Adam, Day (New York: United Nations University, 2020). ISBN: 978-92-808-6513-4.

Finlayson, Andrew R. "A Retrospective on Counterinsurgency Operations: The Tay Ninh Provincial Reconnaissance Unit and Its Role in the Phoenix Program, 1969–70." *Studies in Intelligence* 51(2), 2015: 59–69. Available at: www.cia.gov/library/center-for-the-study-of-intelligence/csi-publications/csi-studies/studies/vol51no2/a-retrospective-on-counterinsurgency-operations.html.

Flor, Leonardo J. "Harnessing Information Operations' Potential Energy." *Military Review*, May–June 2010: 58–64.

Flynn, David. "The Battalion Commander Debates the Blogger (III): I Acted After a Great Deal of Deliberate Planning, Explains LTC Flynn." *Foreign Policy*, 24th January 2011. Available at: https://foreignpolicy.com/2011/01/24/the-battalion-commander-debates-the-blogger-iii-i-acted-after-a-great-deal-of-deliberate-planning-explains-ltc-flynn/.

French, David. *The British Way in Counter-Insurgency 1945–1967* (Oxford: Oxford University Press, 1st Edition, 2011). ISBN: 978-0-19-958796-4.

Fuller, Liz. "Kadyrov's Chechnya Appears Exempt from Russian Funding Cuts." *Radio Free Europe/Radio Liberty*, 30th July 2017. Available at: www.rferl.org/a/caucasus-report-kadyrov-chechnya-exempt-funding-cuts/28648698.html.

Galeotti, Mark. "'Brotherhoods' and 'Associates': Chechen Networks of Crime and Resistance." *Low Intensity Conflict & Law Enforcement* 11(2–3), 2002: 340–352. https://doi.org/10.1080/0966284042000279072.

Galeotti, Mark. *Russia's Wars in Chechnya 1994–2009* (Osprey Publishing, 1st Edition, 2014).

Galula, David. *Counterinsurgency Warfare: Theory and Practice* (London: Praeger Security International, 10th Edition, 2006[1964]). ISBN: 0-275-99269-1.

Garfield, Andrew. "The U.S. Counter-Propaganda Failure in Iraq." *Middle East Quarterly* 14(4), Fall 2007: 23–32. Available at: www.meforum.org/1753/the-us-counter-propaganda-failure-in-iraq.

Garwood, R. "The Second Russo-Chechen Conflict (1999 to Date): 'A Modern Military Operation'?" The *Journal of Slavic Military Studies* 15(3), 2002: 60–103. https://doi.org/10.1080/13518040208430529.

Gilligan, Emma. *Terror in Chechnya: Russia and the Tragedy of Civilians in War* (Princeton: Princeton University Press, 1st Edition, 2010). ISBN: 978-0-691-13079-8.

Girardet, Edward. *Afghanistan: The Soviet War* (Oxon: Routledge, 2011[1985]). ISBN: 978-0-415-68480-4.

Gooch, John. "Re-Conquest and Suppression: Fascist Italy's Pacification of Libya and Ethiopia, 1922–39." *The Journal of Strategic Studies* 28(6), 2005: 1005–1032. https://doi.org/10.1080/01402390500441024.

Gordon, Michael R. "Russia Bombs Chechnya Sites; Major Step-Up." *The New York Times*, 24 September 1999. Available at: www.nytimes.com/1999/09/24/world/russia-bombs-chechnya-sites-major-step-up.html.

Graziani, Rodolfo. *Cirenaica Liberata* (Milano: Mondadori, 1932).

Grebennikov, Marat. "Between Hearts and Minds: The Relevance of the British Colonial Experience to Contemporary Russian Counter-Insurgencies in the North Caucasus." *International Journal* 70(1), 2015: 63–83. https://doi.org/10.1177/0020702014563812.

Gurcan, Metin. *What Went Wrong in Afghanistan? Understanding Counter-Insurgency in Tribalized, Rural and Muslim Environments* (Solihull: Helion and Company Limited, 2016). ISBN: 978-1-911096-00-9.

Hague Convention. "Convention (IV) Respecting the Laws and Customs of War on Land and Its Annex: Regulations Concerning the Laws and Customs of War on Land." *The Hague*, 18th October 1907.

Halbach, Uwe. "Chechnya's Status within the Russian Federation." *German Institute for International and Security Affairs*, SWP Research Paper 2, May 2018.

Herd, Graeme P. "The 'Counter-Terrorist Operation' in Chechnya: 'Information Warfare' Aspects." *Journal of Slavic Military Studies* 13(4), 2000: 57–83. https://doi.org/10.1080/13518040008430460.

Hille, Kathrin. "Chechnya's Economic Recovery Tested by Slowdown." *Financial Times*, 28th April 2015. Available at: www.ft.com/content/8233d33c-ecd0-11e4-a81a-00144feab7de.

Hodgson, Quentin. "Is the Russian Bear Learning? An Operational and Tactical Analysis of the Second Chechen War, 1999–2002." *Journal of Strategic Studies* 26(2), 2003: 64–69. https://doi.org/10.1080/01402390412331302985.

Human Rights Watch. *Rigging the System: Government Policies Co-Opt Aid and Reconstruction in Syria*, June 2019. Available at: www.hrw.org/report/2019/06/28/rigging-system/government-policies-co-opt-aid-and-reconstruction-funding-syria.

Human Rights Watch. *Syria: Residents Blocked from Returning*, 16th October 2018. Available at: www.hrw.org/news/2018/10/16/syria-residents-blocked-returning.

Human Rights Watch. *Targeting Life in Idlib: Syrian and Russian Strikes on Civilian Infrastructure*, October 2020. Available at: www.hrw.org/report/2020/10/15/targeting-life-idlib/syrian-and-russian-strikes-civilian-infrastructure.

Human Rights Watch. *'Welcome to Hell': Arbitrary Detention, Torture, and Extortion in Chechnya*, October 2000. Available at: www.hrw.org/reports/2000/russia_chechnya4/.

Human Rights Watch. *"What Your Children Do Will Touch Upon You": Punitive House-Burning in Chechnya*, July 2009. Available at: www.hrw.org/report/2009/07/02/what-your-children-do-will-touch-upon-you/punitive-house-burning-chechnya.

Hyman, Anthony. "The Struggle for Afghanistan." *The World Today* 40(7), 1984: 276–284.

ICRC (International Committee of the Red Cross). *IV Geneva Convention Relative to the Protection of Civilian Persons in Time of War of 12 August 1949.* Available at: www.un.org/en/genocideprevention/documents/atrocity-crimes/Doc.33_GC-IV-EN.pdf.

ICRC (International Committee of the Red Cross). *Protocol Additional to the Geneva Conventions of 12 August 1949 and Relating to the Protection of Victims of International Armed Conflicts* (Protocol Additional I), 8th June 1977. Available at: www.ohchr.org/en/professionalinterest/pages/protocolii.aspx.

The Independent. *Ramzan Kadyrov: The Warrior King of Chechnya*, 4th January 2007. Available at: www.independent.co.uk/news/people/profiles/ramzan-kadyrov-the-warrior-king-of-chechnya-430738.html.

International Crisis Group. "The Kenyan Military Intervention in Somalia." *Africa Report 184*, 2012.

Jaimoukha, Amjad. *The Chechens: A Handbook* (London: Routledge, 1st Edition, 2005). ISBN: 0-203-35643-8.

The Jamestown Foundation. "Kadyrov Bows Down to Putin . . ." *North Caucasus Weekly* 8(25), 21st June 2007. Available at: https://jamestown.org/program/kadyrov-bows-down-to-putin/.

The Jamestown Foundation. "Kadyrov's Men and the Abuse of Blood Revenge." *North Caucasus Weekly* 4(22), 19 June 2003. Available at: https://jamestown.org/program/kadyrovs-men-and-the-abuse-of-blood-revenge/.

Janeczko, Matthew N. "The Russian Counterinsurgency Operation in Chechnya Part 2: Success, But at What Cost? 1999–2004." *Small Wars Journal*, 2nd November 2012.

Karam, Zeina. "A Luxury City Shows Blueprint for Syria's Rebuilding Plans." *AP News*, 5th November 2018. Available at: https://apnews.com/article/5b5c0055ddfd46d2916f29b9e43349f5.

Kim, Younkyoo & Blank, Stephen. "Insurgency and Counterinsurgency in Russia: Contending Paradigms and Current Perspectives." *Studies in Conflict & Terrorism* 36(11), 2013: 917–932. https://doi.org/10.1080/1057610X.2013.832115.

Kipp, Jacob W. "Russia's Wars in Chechnya." *The Brown Journal of World Affairs* 8(1), 2001: 47–72.

Knight, Amy. *Orders to Kill: The Putin Regime and Political Murder* (New York: St. Martin's Press, 2017). ISBN: 9781250119346.

Knysh, Alexander. "Contextualizing the Salafi-Sufi Conflict (From the Northern Caucasus to Hadramawt)." *Middle Eastern Studies* 43(4), 2007: 503–530. https://doi.org/10.1080/00263200701348847.

Kramer, Mark. "Guerrilla Warfare, Counterinsurgency and Terrorism in the North Caucasus: The Military Dimension of the Russian-Chechen Conflict." *Europe-Asia Studies* 57(2), March 2005: 209–290. https://doi.org/10.1080/09668130500051833.

Kremlin. *Russia Has Never Negotiated with Terrorists*, 6th February 2004. Available at: http://en.kremlin.ru/events/president/news/30315.

Kremlin. *Speech at Expanded Meeting of the State Council on Russia's Development Strategy through to 2020*, 8th February 2008. Available at: http://en.kremlin.ru/events/president/transcripts/24825.

Kurbanova, Majnat. "The Kadyrov System: Neither Russian nor Sharia." *Osservatorio Balcani e Caucaso*, 21st October 2011. Available at: www.balcanicaucaso.org/eng/Areas/Chechnya/The-Kadyrov-system-neither-Russian-nor-sharia-104736.

LandInfo. *Chechnya and Ingushetia: Health Services*, June 2012. Available at: www.landinfo.no/asset/2322/1/2322_1.pdf.

Lapidus, Gail W. "Contested Sovereignty: The Tragedy of Chechnya." *International Security* 23(1), Summer 1998: 5–49. Available at: www.jstor.org/stable/2539261.

Lyall, Jason. "Are Coethnics More Effective Counterinsurgents? Evidence from the Second Chechen War." *The American Political Science Review* 104(1), 2010: 1–20. Available at: www.jstor.org/stable/27798537.

MACV. "Captured VCI Documents." *Declassified Documents*, 25th August 1970b. Available at: https://archive.org/details/Captured-VCI-docs-RE_-Phoenix-Program-Aug-Dec-1970.

MACV. "Phung Hoang SOP." *Declassified Documents*, 1970a. Available at: https://archive.org/details/Phoenix-Standard-Operating-Procedure-Feb-70/page/n12.

Martínez, Ciro José & Eng, Brent. "Struggling to Perform the State: The Politics of Bread in the Syrian Civil War." *International Political Sociology* 11, 2017: 130–147. https://doi.org/10.1093/ips/olw026.

Matveeva, Anna. "Chechnya: Dynamics of War and Peace." *Problems of Post-Communism* 54(3), 2007: 3–17. https://doi.org/10.2753/PPC1075-8216540301.

Meakins, Joss. "The Other Side of the COIN: The Russians in Chechnya." *Small Wars Journal*, 15th January 2017. Available at: https://smallwarsjournal.com/jrnl/art/the-other-side-of-the-coin-the-russians-in-chechnya.

Metz, Thomas F.; Garrett, Mark W., Hutton, James E. & Bush, Timothy W. "Massing Effects in the Information Domain: A Case-Study in Aggressive Information Operations." *Military Review*, May–June 2006: 2–12.

Meyerfeld, Bruno. "Somalia: Madobe, The Respectable Jihadist." *True Story Award*, 2019. Available at: https://truestoryaward.org/story/44.

Miakinkov, Eugene. "The Agency of Force in Asymmetrical Warfare and Counterinsurgency: The Case of Chechnya." *Journal of Strategic Studies* 34(5), 2011: 647–680. https://doi.org/10.1080/01402390.2011.608946.

Mitrokhin, Vasily. "The KGB in Afghanistan." *Woodrow Wilson International Center for Scholars*, Working Paper 40, 2002. Available at: www.wilsoncenter.org/publication/the-kgb-afghanistan.

Moe, Louise W. "Counterinsurgent Warfare and the Decentering of Sovereignty in Somalia." In *Reconfiguring Intervention: Complexity, Resilience and the 'Local*

Turn' in Counterinsurgent Warfare, edited by Moe, Louise W. & Müller, Markus-Michael (London: Palgrave Macmillan, 2017). https://doi.org/10.1057/978-1-137-58877-7-6.

Molan, Jim. "Do You Need to Decisively Win the Information War? Managing Information on Operations in Iraq." *Security Challenges* 5(1), 2009: 37–52.

Moore, Cerwyn & Tumelty, Paul. "Assessing Unholy Alliances in Chechnya: From Communism and Nationalism to Islamism and Salafism." *Journal of Communist Studies and Transition Politics* 25(1), 2009: 73–94. https://doi.org/10.1080/13523270802655621.

Mutambo, Aggrey. "Is Madobe Ringfencing Jubbaland Presidency?" *The East African*, 20th July 2019. Available at: www.theeastafrican.co.ke/tea/news/east-africa/is-madobe-ringfencing-jubbaland-presidency—1423008.

Myers, Steven L. "Chechnya Bomb Kills President, A Blow to Putin." *The New York Times*, 10th May 2004. Available at: www.nytimes.com/2004/05/10/world/chechnya-bomb-kills-president-a-blow-to-putin.html.

Newsinger, John. *British Counterinsurgency: From Palestine to Northern Ireland* (Houndmills: Palgrave Macmillan, 2002). ISBN: 978-1-349-41996-8.

Office for the Coordination of Humanitarian Affairs (OCHA). "Endless Brutality: Ongoing Human Rights Violations in Chechnya." *Physicians for Human Rights*, 23rd January 2001. Available at: https://reliefweb.int/report/russian-federation/endless-brutality-ongoing-human-rights-violations-chechnya.

Oliker, Olga. *Russia's Chechen Wars 1994–2000: Lessons from Urban Combat* (Santa Monica: RAND Corporation, 1st Edition, 2001). ISBN: 0-8330-2998-3.

O'Loughlin, John & Witmer, Frank D. W. "The Diffusion of Violence in the North Caucasus of Russia, 1999–2010." *Environment and Planning A* 44, 2012: 2379–2396. https://doi.org/10.1068/a44366.

Osborn, Andrew & Solovyov, Dimitry. "Chechen Leader, amid Reshuffles, Says Ready to Die for Putin." *Reuters*, 27th November 2017. Available at: www.reuters.com/article/us-russia-chechnya/chechen-leader-amid-reshuffles-says-ready-to-die-for-putin-idUSKBN1DR03I.

Owens, Patricia. *Economy of Force: Counterinsurgency and the Historical Rise of the Social* (Cambridge: Cambridge University Press, 2015). ISBN: 978-1-107-12194-2.

Pain, Emil. "From the First Chechen War Towards the Second." *The Brown Journal of World Affairs* 8(1), 2001: 7–19.

Peterson, Scott. "Revenge Fuels Chechen Flames." *The Christian Science Monitor*, 8th October 2003. Available at: www.csmonitor.com/2003/1008/p06s01-woeu.html.

Pictet, Jean S. *The Geneva Conventions of 12 August 1949: Commentary, Volume IV Relative to the Protection of Civilian Persons in Time of War* (Geneva: International Committee of the Red Cross, 1958). ISBN: 90-247-3460-6.

Politkovskaya, Anna. *A Small Corner of Hell: Dispatches from Chechnya* (Chicago: The University of Chicago Press, 5th Edition, 2003). ISBN: 0-226-67432-0.

Radio Free Europe/Radio Liberty. *Russia: Chechen Official Puts War Death Toll at 160,000*, 16th August 2005. Available at: www.rferl.org/a/1060708.html.

Rahmani, Abdul R. & Lawrence, Noor A. "How Do We Win Information Warfare in Afghanistan?" *Small Wars Journal*, 2018. Available at: https://smallwarsjournal. com/jrnl/art/how-do-we-win-information-warfare-afghanistan.

Ratelle, Jean-François. "North Caucasian Foreign Fighters in Syria and Iraq: Assessing the Threat of Returnees to the Russian Federation." *Caucasus Survey* 4(3), 2016: 218–238. https://doi.org/10.1080/23761199.2016.1234096.

Rizvi, Naseem. "Sovietization of Afghan Society." *Strategic Studies* 12(1), 1988: 78–100. Available at: www.jstor.org/stable/45182763.

Robinson, Paul. "Soviet Hearts and Minds Operations in Afghanistan." *The Historian* 72(1), 2010: 1–22.

Russell, John. "Chechen Elites: Control, Cooption, or Substitution?" *Europe-Asia Studies* 63(6), 2011a: 1073–1087. https://doi.org/10.1080/09668136.2011.585758.

Russell, John. "Kadyrov's Chechnya—Template, Test or Trouble for Russia's Regional Policy?" *Europe-Asia Studies* 63(3), 2011b: 509–528. https://doi.org/10 .1080/09668136.2011.557541.

Russell, John. "Ramzan Kadyrov: The Indigenous Key to Success in Putin's Chechenization Strategy?" *Nationalities Papers* 36(4), 2008: 659–687. https://doi. org/10.1080/00905990802230605.

SAMS. *Two Medical Staff Killed in Russian Airstrikes on Sarmin*, 22nd October 2015. Available at: www.sams-usa.net/press_release/press-release-two-medical-staff-killed-in-russian-airstrikes-on-sarmin/.

Schaefer, Robert W. *The Insurgency in Chechnya and the North Caucasus: From Gazavat to Jihad* (Santa Barbara: Praeger Security International, 2010). ISBN: 978-0-313-38634-3.

Schmitt, Michael N. "State-Sponsored Assassination in International and Domestic Law." *Yale Journal of International Law* 17(609), 1992: 652–684.

Seierstad, Åsne. *The Angel of Grozny: Orphans of a Forgotten War* (New York: Basic Books, 1st Edition, 2008). ISBN-13: 978-0-465-01122-3.

Smirnov, Andrei. "The Kremlin's New Strategy to Build a Pro-Russian Islamic Chechnya." *North Caucasus Weekly* 7(9), The Jamestown Foundation 2006. Available at: https://jamestown.org/program/the-kremlins-new-strategy-to-build-a-pro-russian-islamic-chechnya-2/.

Souleimanov, Emil A. "Attacks in Chechnya Suggest Opposition to Kadyrov is Far from Eradicated." *The Central Asia-Caucasus Analyst*, 24th March 2017b. Available at: https://cacianalyst.org/publications/analytical-articles/item/13436-attacks-in-chechnya-suggest-opposition-to-kadyrov-is-far-from-eradicated.html.

Souleimanov, Emil A. "An Ethnography of Counterinsurgency: Kadyrovtsy and Russia's Policy of Chechenization." *Post-Soviet Affairs* 31(2), 2015: 91–114. https://doi.org/10.1080/1060586X.2014.900976.

Souleimanov, Emil A. *The North Caucasus Insurgency: Dead or Alive?* (Carlisle Barracks: Strategic Studies Institute, U.S. Army War College, February 2017a). ISBN: 1-58487-748-0.

Souleimanov, Emil A. "Russian Chechnya Policy: 'Chechenization' Turning into 'Kadyrovization'?" *The Central Asia-Caucasus*, 31st May 2006: 3–5.

Souleimanov, Emil A.; Abbasov, Namig & Siroky, David S. "Frankenstein in Grozny: Vertical and Horizontal Cracks in the Foundation of Kadyrov's Rule." *Asia*

Europe Journal 17(87), 2019: 87–103. https://doi.org/10.1007/s10308-018-0520-y.

Souleimanov, Emil A. & Aliyev, Huseyn. "Asymmetry of Values, Indigenous Forces, and Incumbent Success in Counterinsurgency: Evidence from Chechnya." *Journal of Strategic Studies* 38(5), 2015a: 678–703. https://doi.org/10.1080/0140239 0.2014.952409.

Souleimanov, Emil A. & Aliyev, Huseyn. "Blood Revenge and Violent Mobilization: Evidence from the Chechen Wars." *International Security* 40(2), 2015b: 158–180. https://doi.org/10.1162/ISEC_a_00219.

Souleimanov, Emil A. & Aliyev, Huseyn. *How Socio-Cultural Codes Shaped Violent Mobilisation and Pro-insurgent Support in the Chechen Wars* (Cham: Palgrave Macmillan, 2017). https://doi.org/10.1007/978-3-319-52917-2.

Souleimanov, Emil A. & Jasutis, Grazvydas. "The Dynamics of Kadyrov's Regime: Between Autonomy and Dependence." *Caucasus Survey* 4(2), 2016: 115–128. https://doi.org/10.1080/23761199.2016.1183396.

Souleimanov, Emil A. & Siroky, David S. "Random or Retributive? Indiscriminate Violence in the Chechen Wars." *World Politics* 68(4), 2016: 677–712. https://doi.org/10.1017/S0043887116000101.

Sperotto, Federico. "Counter-Insurgency, Human Rights, and the Law of Armed Conflict." *Human Rights Brief* 17(1), 2009: 19–23.

Steele, Jonathan. "It Is Over, and Putin Won." *The Guardian*, 30th September 2008. Available at: www.theguardian.com/commentisfree/2008/sep/30/russia.chechnya.

Strawser, Bradley J. "Moral Predators: The Duty to Employ Uninhabited Aerial Vehicles." *Journal of Military Ethics* 9(4), 2010: 342–368. https://doi.org/10.108 0/15027570.2010.536403.

Takriti, Abdel R. *Monsoon Revolution: Republicans, Sultans, and Empires in Oman, 1965–1976* (Oxford: Oxford University Press, 2013), 200. ISBN: 978-0-19-967443-5.

Tchantouridzé, Lasha. "Counterinsurgency in Afghanistan: Comparing Canadian and Soviet Efforts." *International Journal* 68(2): 331–345. https://doi.org/10.1177/0020702013492517.

Thomas, Timothy L. "Information Warfare in the Second (1999–) Chechen War. Motivation for Military Reform?" In *Russian Military Reform 1992–2002*, edited by Aldis, Anne C. & McDermott, Roger N. (London: Frank Cass, 2003). ISBN: 0-7146-5475-2.

Thomas, Timothy L. "Russian Tactical Lessons Learned Fighting Chechen Separatists." *Journal of Slavic Military Studies* 18(4), 2005: 731–766. https://doi.org/10.1080/13518040500355015.

Toft, Monica D. & Zhukov, Yuri M. "Islamists and Nationalists: Rebel Motivation and Counterinsurgency in Russia's North Caucasus." *American Political Science Association* 109(2), 2015: 222–238. https://doi.org/10.1017/S000305541500012X.

Trentin, Dimitri V. & Malashenko, Aleksei V. *Russia's Restless Frontier: The Chechnya Factor in Post-Soviet Russia* (Washington, DC: Carnegie Endowment for International Peace, 2004). ISBN: 0-87003-203-8.

Trinquer, Robert. *Modern Warfare: A French View of Counterinsurgency* (Fort Leavenworth: U.S. Army Command and General Staff College, 1985).

United Nations (U.N.). *SRGS Kay's Speech on the Occasion of the Formal Inauguration of the Interim Jubba Administration*, 21st January 2014. Available at: https://unsom.unmissions.org/srsg-kays-speech-occasion-formal-inauguration-interim-jubba-administration.

U.S. Government. *The Commander's Handbook on the Law of Naval Operations* (Norfolk: Department of the Navy, August 2017).

U.S. Government. *Soviet Influence Activities: A Report on Active Measures and Propaganda, 1986–87* (Washington, DC, August 1987).

Walsh, Nick P. & Habib, Nasir. "Pakistani Leaders Condemn Suspected U.S. Drone Strike." *CNN*, 21st March 2011. Available at: http://edition.cnn.com/2011/WORLD/asiapcf/03/18/pakistan.drone.strike/?hpt=T2.

Ware, Robert B. "Chechenization: Ironies and Intricacies." *Brown Journal of World Affairs* 15(2), 2009: 157–169.

Wingfield, Thomas C. "International Law and Information Operations." In *Cyberpower and National Security*, edited by Kramer, Franklin D.; Starr, Stuart H. & Wentz, Larry K. (Washington, DC: Potomac Books, 2009). ISBN: 978-1-59797-423-3.

World Peace Foundation. *Mass Atrocity Index. Russia: Chechen War*, 2015. Available at: https://sites.tufts.edu/atrocityendings/2015/08/07/russia-1st-chechen-war/.

5   Conclusion

What makes a counterinsurgent successful? The conventional wisdom on COIN warfare suggests that a COIN campaign ends up in failure if the counterinsurgent does not gain the population's trust, cannot minimise the use of force, or is unwilling to solve the popular grievances on which the rebels feed on to gain resources and attract new recruits. The hearts and minds approach has long been considered a universal textbook for success, a formula that promises the attainment of victory by putting the well-being of non-combatants ahead of any other strategic consideration.

Yet, hearts and minds are not the only game in town. Counterinsurgents often resort to brutalisation-centred measures in an effort to quell the insurgency and cow the local population into submission. Seen as a counterproductive and morally appalling way of fighting insurgency, the brutalisation paradigm has long escaped the scholarship's scrutiny. Little is known about the mechanisms and logic of brutalisation nor do we possess a solid understanding of whether coercive measures can deliver mission success. Although brutalisation-centred measures have found widespread application in past and present theatres of asymmetric warfare, their highly controversial nature has kept the scholarship away from exploring this long-neglected variant of COIN warfare. While critics of Western COIN blueprints have exposed the shortfalls affecting the hearts and minds approach, their writings have paid marginal attention to the lessons that practitioners can draw from their findings. Western decision-makers are aware of the potential strategic setbacks brought about by the hearts and minds approach, yet they know little about the alternatives available to counterinsurgents struggling to win over the population's support. This book has overcome the psychological barrier placed against the study of controversial COIN techniques and expounded why, when, and how brutalisation-centred measures can break an insurgency's backbone and deter the population from engaging in the anti-incumbent activity. We argued that counterinsurgents opting for the brutalisation paradigm follow a "shortcut approach" designed to bypass the

DOI: 10.4324/9781003188049-5

moral and ethical restrictions typical of hearts and minds operations. Engaging in theoretical debates on the nature of COIN warfare and drawing upon an extensive overview of empirical evidence, the book revealed that, under certain conditions, brutalisation can match, if not outperform the hearts and minds approach. When following the brutalisation paradigm, counterinsurgents exploit intra-communal antagonisms to gather intelligence, employ IO to dehumanise their opponents, victimise civilians to deter non-compliance, co-opt indigenous elites to engender legitimacy, and weaponise economic inducements to minimise the incidence of anti-incumbent activity. We have found that brutalisation-centred measures provide counterinsurgents with an effective toolkit designed to swiftly suppress grassroots insurgencies and deter the population from engaging in anti-incumbent activities.

The rigorous scrutiny of empirical data from the Russian COIN operations in Chechnya and elsewhere provides solid confirmation to the aforementioned theoretical findings. By resorting to a well-balanced combination of brutalisation-centred measures, Moscow successfully crushed the rampant Chechen insurgency and regained control over the small breakaway republic. If waging a campaign of indiscriminate violence enabled the Russians to rapidly seize and maintain territorial control, setting up a tight network of local informants capable of submitting timely and actionable intelligence allowed the pro-Russian camp to effectively stem the tide of people joining the rebellion. Acting upon increasingly accurate intelligence submitted by turncoat rebels and local informants, the Russians selectively targeted suspected insurgents and punished their families for the crime of defying the incumbent. Instead of allowing the population to choose a suitable leader of its own free will, Moscow exacerbated intra-clan feuds to "indigenise" the conflict and endorse a loyal leadership capable of garnering genuine support among local key actors. The weaponisation of economic inducements further inflicted a striking blow on the insurgents, who could no longer find shelter and assistance amongst communities all too aware of the punishments awaiting those who dared defying Kadyrov's regime. With most of its fighters either killed or dissuaded from further action, its support network largely dismantled by the security forces, and its popular mass base deterred into passive acquiesce, the insurgency progressively collapsed under the pressure exercised by the "rule-by-fear machine" set up by the pro-Russian regime (Souleimanov et al. 2019: 98). At the time of writing, more than 10 years after the official end of the COIN operations, the situation in Chechnya remains relatively stable. Against a backdrop of widespread scepticism, the "Russian shortcut" seems to have ticked all boxes in the checklist of a successful COIN campaign.

These findings bear important implications for Western practitioners at a time of deep crisis for the hearts and minds approach. Despite thousands of lives sacrificed and billions of dollars dissipated in striving to win over the local populations' wholehearted support, results obtained by Western counterinsurgents have been "distinctively unimpressive" at best and utterly disastrous at worst (Gray 2013: vii). While rebel groups continue inflicting heavy sufferings upon the Iraqi population, killing more than 2.300 civilians in 2019 alone, the situation in Afghanistan is worsening as the government loses ground to a resilient insurgency controlling almost 20% of the national territory (Iraq Body Count 2020; Roggio & Gutowski 2020). As these setbacks call for a re-thinking of the precepts encompassed in Western military manuals, looking at the experience of successful counterinsurgents could offer new solutions for the shortfalls of the hearts and minds approach. While this book strongly discourages Western practitioners from replicating Russia's patently criminal strategies, its findings suggest that a COIN force can hardly be victorious if it is not willing to exploit the full range of available measures at its disposal.

It goes without saying that discarding important moral and legal principles for the conduct of COIN warfare is impermissible and unacceptable for a Western military force. Legal frameworks are instrumental for developing coherent sets of strategic, operational, and tactical measures in support of effective hearts and minds endeavours. Brutalising the population might appear to some as an easy way out of a conflict with no end in sight; yet, conceding to these impulses would be a fatal mistake. As noted by Galvin decades before the U.S. involvement in Afghanistan and Iraq, the hearts and minds approach can hardly succeed if the counterinsurgent fails to take into account the imperative of shielding the population from unnecessary sufferings: "If, for example, the military's actions in killing 50 guerrillas cause 200 previously uncommitted citizens to join the insurgent cause, the use of force will have been counterproductive" (Galvin 1986: 6). That said, such considerations should not discourage military commanders from taking full advantage of the strategic options at their disposal within the boundaries of legality. According to IHL principles, the military necessity of suppressing insurgency justifies the proportionate use of coercive measures, and a responsible counterinsurgent should be able to decide when a more assertive use of force is necessary to achieve key operational objectives. As stated by Sloane, decision-makers should consider the LOAC as an asset, not as an impediment:

> Any proposed LOAC rule or principle must enable the reasonable military commander to observe the law's posited limits in good faith and simultaneously be confident of his ability to pursue effectively the military goals with which, in a liberal polity committed to the rule of law, he has been charged by those on whose behalf the military acts.
>
> (Sloane 2015: 334)

It is within these legal boundaries that the takeaways advanced in this book are to be intended. By harmonising the principles encompassed in hearts and minds blueprints with a more assertive use of coercive measures in accordance with LOAC principles, Western counterinsurgents could better contain the spread of rebel activity without infringing the moral requirements set forth in their operational guidelines.

This book has also revealed that the mechanisms of brutalisation in COIN warfare remain largely unknown and widely misunderstood. Calling for further research into this long-overlooked paradigm, this book demonstrated that diving deep into alternative strategic approaches discloses important implications for the theory and practice of COIN warfare. As a last thought for future researchers, we recommend to closely monitor the dynamics of brutalisation endeavours carried out in urbanised environments, such as the ones that have characterised the Syrian Civil War. With the rapid urbanisation of the Third World pushing large masses of impoverished people to leave the countryside for large metropolises, the contextual settings of future COIN operations appear less like Chechnya's woods-covered peaks and more like the smoking ruins of Aleppo's suburbs. With the "coming age of the urban guerrilla" rapidly approaching, we join Kilcullen in arguing that the time has come for researchers to "drag ourselves—body and mind—out of the mountains" (Kilcullen 2013: 262).

In rationalising the logic of brutalisation COIN campaigns, this study has not aimed at justifying the use of torture, forced starvation, mass killing, and other repulsive techniques of warfare. Its purpose was to elucidate the reasons why these supposedly counterproductive methods display a surprisingly high record of successes. Understanding—and therefore predicting—which techniques are more effective than others in suppressing insurgency is of the utmost importance for present and future practitioners. This book revealed that brutalisation-centred measures can and do deliver success in the fight against grassroots insurgencies. What remains to be established is whether a marriage between coercive and persuasive measures could put an end to the trail of disappointing results achieved by Western states in the last two decades of uninterrupted COIN operations.

References

Galvin, John R. "Uncomfortable Wars: Toward a New Paradigm." *Parameters* 16(4), 1986: 6.

Gray, Colin S. "Foreword." In *Counterinsurgency in Crisis: Britain and The Challenges of Modern Warfare*, edited by Ucko, David H. & Egnell, Robert (New York: Columbia University Press, 2013). Available at: www.jstor.org/stable/10.7312/ucko16426.3.

Iraq Body Count. *Documented Civilian Deaths from Violence.* Accessed April 2020. Available at: www.iraqbodycount.org/database/.

Kilcullen, David. *Out of the Mountains: The Coming Age of the Urban Guerrilla* (London: Hurst, 1st Edition, 2013). ISBN: 978-0-19-973750-5.

Roggio, Bill & Gutowski, Alexandra. "Mapping Taliban Control in Afghanistan." *FDD's Long War Journal.* Accessed April 2020. Available at: www.longwarjournal.org/mapping-taliban-control-in-afghanistan.

Sloane, Robert D. "Puzzles of Proportion and the 'Reasonable Military Commander': Reflections on the Law, Ethics, and Geopolitics of Proportionality." *Harvard National Security Journal* 6, 2015: 299–343.

Souleimanov, Emil A.; Abbasov, Namig & Siroky, David S. "Frankenstein in Grozny: Vertical and Horizontal Cracks in the Foundation of Kadyrov's Rule." *Asia Europe Journal* 17(87), 2019: 87–103. https://doi.org/10.1007/s10308-018-0520-y.

Index